INSTANT REFERENCE ASTRONOMY

 TEACH YOURSELF®

For UK orders: please contact Bookpoint Ltd, 78 Milton Park, Abingdon, Oxon OX14 4TD. Telephone: (44) 01235 400414, Fax: (44) 01235 400454. Lines are open 9.00-6.00, Monday to Saturday, with a 24-hour message answering service. E-mail address: orders@bookpoint.co.uk

For USA and Canada orders: please contact NTC/Contemporary Publishing, 4255 West Touhy Avenue, Lincolnwood, Illinois 60646-1975, USA. Telephone: (847) 679 5500, Fax: (847) 679 2494.

Long renowned as the authoritative source for self-guided learning — with more than 40 million copies sold worldwide — the *Teach Yourself* series includes over 200 titles in the fields of languages, crafts, hobbies, business, computing and education.

British Library Cataloguing in Publication Data
A catalogue record for this title is available from the British Library.

Library of Congress Catalog Card Number: On file

First published in UK 2000 by Hodder Headline Plc, 338 Euston Road, London NW1 3BH.

First published in US by NTC/Contemporary Publishing, 4255 West Touhy Avenue, Lincolnwood (Chicago), Illinois 60646-1975, USA.

The 'Teach Yourself' name and logo are registered trademarks of Hodder & Stoughton.

Picture credits: Ann Ronan at Image Select 6, 22, 121, Ann Ronan Picture Library 37, 38, 55, 69, 83, 119, 120, 140, Image Select 20, 33, 36, 147, 171, 181, Image Select/Ann Ronan 32, NASA/Image Select 51, 125.

Text editor: Christopher Cooper
Typeset by TechType, Abingdon, Oxon
Printed in Great Britain for Hodder & Stoughton Educational, a division of Hodder Headline Plc, 338 Euston Road, London NW1 3BH, by Cox & Wyman Ltd, Reading, Berkshire.

Impression number   10 9 8 7 6 5 4 3 2 1
Year   2006 2005 2004 2003 2002 2001 2000

Contents

Bold type in the text indicates a cross reference. A plural, or possessive, is given as the cross reference, i.e. is in bold type, even if the entry to which it refers is singular.

aberration of starlight

Stars seem to change in position because of the motion of the Earth around the Sun. The Earth's movement makes the starlight seem to come from a slightly different direction, just as rain that is really falling vertically seems to fall at an angle when seen from a moving train or car. The effect was discovered by James Bradley (1693–1762) in 1729 and is direct evidence for the movement of the Earth around the Sun.

absorption lines

Dark lines in the **spectrum** of a star, **nebula,** or other astronomical object, caused by gas along the line of sight. Atoms in the gas absorb light from the source at particular wavelengths. Numerous absorption lines in the **spectrum** of the Sun (**Fraunhofer lines**) allow astronomers to study the composition of the Sun's outer layers. Absorption lines in the spectra of stars give clues to the composition of interstellar gas.

rain falling past window of stationary train

rain falling past window of moving train

direction in which star seems to lie

true position of star

starlight enters telescope

starlight reaches eyepiece

movement of Earth

aberration of starlight *The aberration of starlight is an optical illusion caused by the motion of the Earth. Light from a star 'falling' down a telescope seems to follow a sloping path because the Earth is moving. This causes an apparent displacement, or aberration, in the position of the star.*

accretion

The process by which an astronomical object, such as a **planet** or star, gathers material from the surrounding space by gravitational attraction. The object increases in mass and releases gravitational energy as the material falls onto it.

   See also: *gravity.*

Achernar or Alpha Eridani
The brightest star in the **constella-tion** Eridanus, and the ninth-brightest star in the sky. It is a hot, luminous blue star with a true **luminosity** 250 times that of the Sun. Achernar is 125 **light years** away from the Sun.

Achilles group
A group of **asteroids** that share the **orbit** of **Jupiter**. The asteroids move around a point 60° ahead of the **planet**.
   See also: *Patroclus group, Trojan asteroids.*

Accretion on to compact objects such as white dwarfs, neutron stars, and black holes releases huge amounts of gravitational energy, and is believed to be the power source for active galaxies and quasars. Accreted material falling towards a star may form a swirling disc of material known as an accretion disc that can be a source of X-rays.

achondrite
A type of **meteorite**, the remains of a piece of rock that has fallen to Earth from space. About 15% of all meteorites are achondrites. They lack the **chondrules** (silicate spheres) found in **chondrites**.

Acrux or Alpha Crucis
The brightest **star** in the **constellation** of **Crux**, the Southern Cross, and the 12th-brightest star in the night sky. It is a **binary star** comprising two blue-white stars and is 510 **light years** away. Together with nearby Gacrux, it points towards the south celestial pole.

active galaxy
A type of **galaxy** that emits vast quantities of **light**, radio waves, **X-rays** and other forms of energy from a small region at its centre called the active galactic nucleus (AGN). Active galaxies are subdivided into radio galaxies, **Seyfert galaxies**, BL Lacertae objects, and **quasars**. They are thought to con-tain **black holes** with a mass as much as 100 million times that of the Sun, drawing stars and interstellar gas towards it in a process of accretion. The gravitational energy released by the infalling material is the power source for the AGN. Some of the energy may appear as a pair of jets of hot gas emerging in opposite direc-tions from the nucleus.

The differences between the different sorts of active galaxy may be simply due to the way we look at them. The orientation of the jets of hot gas to the line of sight and their interaction with surrounding material determines the type of object that we see.

airglow

Faint and variable **light** in the Earth's **atmosphere** produced by chemical reactions in the **ionosphere**. The reactions are the recombinations of ions, fragments of atoms produced by the Sun's radiations.

albedo

The fraction of the incoming **light** reflected by a body such as a **planet**. A body with a high albedo, near 1, is very bright, while a body with a low albedo, near 0, is dark.

Albedos of planets in our Galaxy

Object	Albedo	Object	Albedo
Mercury	0.10	Saturn	0.47
Venus	0.60	Uranus	0.57
Earth	0.37	Neptune	0.51
Mars	0.15	Pluto	0.12
Jupiter	0.44		

Aldebaran or Alpha Tauri

The brightest star in the **constellation Taurus** (the Bull) and the 14th-brightest star in the night sky. It marks the eye of the bull. Aldebaran is a **red giant** 60 **light years** away from the Earth, shining with a true luminosity of about 100 times that of the Sun.

In ancient Persia, Aldebaran was the first of four Royal Stars that marked the approximate positions of the Sun at the time of the **equinoxes** and **solstices**. The other three were **Regulus**, **Antares**, and **Fomalhaut**. The Romans knew it as 'Palilicium' because it disappeared into the evening twilight about the time when the feast of Pales, the Roman deity of shepherds and their flocks, was celebrated.

Algol or Beta Persei

A **variable** star that consists of two stars orbiting each other, in the **constellation Perseus**. The fainter star **eclipses** the brighter one every 69 hours, causing the apparent brightness of the pair to drop by two-thirds. Between these events, the brighter star eclipses the fainter one, causing a less noticeable dip in brightness.

UNDERSTANDING ALGOL

Algol's brightness changes were first explained in 1782 by an English amateur astronomer, John Goodricke (1764–1786). He pointed out that the changes between magnitudes 2.2 and 3.5 repeated themselves exactly after an interval of 2.867 days and supposed this to be due to two stars orbiting round and eclipsing each other. This was the first eclipsing binary to be recognized. Later, more refined observations showed a third, fainter star, revolving round the brighter pair in 1.87 years.

Alpha Centauri or Rigil Kent
The brightest star in the **constellation Centaurus** and the third-brightest star in the night sky. It is actually a triple star: the two brighter stars **orbit** each other every 80 years. The third, **Proxima Centauri**, is the closest star to the Sun, 4.2 **light years** away, 0.1 light years closer than the other two.

alpha particle
A subatomic particle produced in some types of radioactivity and other sub-atomic interactions. It is identical to the nucleus of an atom of ordinary **helium** and plays an important part in the reactions that generate heat and **light** inside stars and the Sun.

Altair or Alpha Aquiline
The brightest star in the **constellation Aquila** and the 13th-brightest star in the night sky. It is a white star 16 **light years** away from the Sun and forms the Summer Triangle with the stars **Deneb** (in the constellation **Cygnus**) and **Vega** (in **Lyra**).

altazimuth
An astronomical instrument designed for observing the **altitude** and **azimuth** of a celestial object.

altitude or elevation
The angle between an object in the sky and the horizon. It ranges from 0° on the horizon to 90° at the **zenith**. Together with **azimuth**, it forms the system of horizontal coordinates for specifying the positions of celestial bodies.

amateur astronomy
Hundreds of thousands of people who are not professional astronomers enjoy studying the sky and contribute to scientific knowledge of celestial objects. National and local societies bring amateur astronomers together

with each other and with professionals. Amateurs can pursue astronomy in many ways:

- Even with the naked eye one can follow the changing appearance of the stars and the movements of the Moon and planets through the year. Serious study of the rates of **meteor showers** is possible.

- Amateurs can also make impressive photographs of the sky with ordinary cameras.

- With **binoculars** or small **telescopes** amateurs can observe **binary stars**, star clusters, **nebulae,** and **galaxies** that are invisible to the naked eye.

- Amateurs can track the changes in brightness of variable stars, and the changing appearance of **Mars**, **Jupiter**, **Saturn,** and their **satellites**.

- Most new **comets** are discovered by dedicated amateurs who sweep the skies with **binoculars** on every clear night.

- The most sophisticated amateur astronomers use large telescopes linked to electronic detectors and computers, and some use spectroscopes and **radio telescopes**.

Amor asteroids
A group of **asteroids** that sometimes come close to the Earth. Their orbits lie completely outside the Earth's **orbit** but intersect the orbit of **Mars**. Some have probably collided with Mars in the past, and some could collide with the Earth in the future.

Andromeda
A major **constellation** of the northern hemisphere, high in the night sky in autumn. Its main feature is the **Andromeda galaxy**. The star Alpha Deneb, together with the three brightest stars of the constellation **Pegasus**, forms the Square of Pegasus.

Andromeda is named after an Ethiopian princess of Greek mythology, who was chained to a rock as a sacrifice to a sea monster. She was rescued by Perseus, who married her. Her father Cepheus, her mother Cassiopeia, and the monster Cetus, are nearby in the sky.

Andromeda galaxy
A **galaxy** 2.2 million **light years** away from Earth in the constellation **Andromeda**, and the most distant object visible to the naked eye. It is the largest member of the **Local Group** of galaxies. Like our galaxy, the **Milky Way**, it is a spiral orbited by several companion galaxies, but it contains about twice as many stars as the Milky Way. It is about 200,000 light years across.

In 1993 US astronomers detected two components at the centre of Andromeda, indicating that it may have a double nucleus. The fainter of the two components lies at the exact centre of Andromeda while the rest of the galaxy orbits it. A **black hole** with a mass tens of millions times that of the Sun has been detected here. The other, brighter component of the 'double nucleus' may be a remnant of a galaxy that collided with Andromeda.

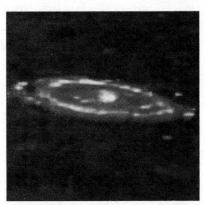

Andromeda Galaxy *This image of the Andromeda Galaxy was made by the Infrared Astronomical Satellite (IRAS).*

angular momentum

An important property of an object that is rotating (such as a wheel) or revolving around another one (such as a **planet** moving around the Sun). A wheel, for example, gains angular momentum if it spins faster. If two wheels are spinning at the same number of revolutions per minute, one will have a greater angular momentum if it has more mass (while being the same size and shape); or if it has a bigger radius (while having the same mass). In an isolated system angular momentum does not change.

Angular momentum is crucial to many astronomical processes:

- In a cloud of gas that shrinks to form a star or planet, the cloud has to spin faster and faster as it shrinks (as ice-skaters spin faster when they pull in their arms).

- In the Earth–Moon system, the drag of the **tides**, attracted by the Moon's gravity, slows the Earth's rotation. The angular momentum lost from the Earth is gained by the Moon, which is slowly spiralling away from the Earth in consequence.

annular eclipse

A **solar eclipse** in which the Moon appears slightly smaller than the Sun and a thin ring of sunlight remains visible around the edge of the Moon. Annular eclipses occur when the Moon is at its farthest point from the Earth.

See also: *partial eclipse, total eclipse.*

Antares or Alpha Scorpio

The brightest star in the **constellation Scorpius** (the scorpion) and the 15th-brightest star in the night sky. It is a red **supergiant** several hundred times larger than the Sun and perhaps 10,000 times as luminous. It lies about 300 **light years** away from the Sun, and varies in brightness.

antimatter

A form of matter that has the opposite properties to ordinary matter. Most of the attributes of elementary particles (such as electric charge, magnetic moment, and spin) are reversed. Particles of antimatter, called antiparticles, can be created in particle accelerators, such as those at CERN in, Switzerland, and at Fermilab in the USA.

INSTANT ANTIMATTER

In 1996 physicists at CERN created whole atoms of antimatter: nine atoms of antihydrogen survived for 40 nanoseconds (40 billionths of a second).

A large amount of antimatter was probably created at the birth of the universe in the **Big Bang**, but was immediately annihilated in collisions with ordinary matter. There was a slight excess of ordinary matter left over, which makes up the universe today.

apastron

The point at which an object travelling in an elliptical orbit around a star is furthest from the star; the point at which it is nearest is **periastron**.

aperture synthesis

A technique used in **radio astronomy** in which several small radio dishes are linked to simulate the performance of one very large **radio telescope** many kilometres in diameter.

apex

The point in the sky towards which the Sun and Solar System are travelling, because of their movement around the centre of our **Galaxy**. It lies in the constellation **Hercules**.

aphelion

The point at which an object, travelling in an elliptical **orbit** around the Sun, is at its farthest from the Sun. The Earth is at its aphelion on 5 July.

apogee

The point at which an object, travelling in an elliptical **orbit** around the Earth, is at its farthest from the Earth.

Apollo asteroids

A group of **asteroids** whose **orbits** cross that of the Earth. They are named after the first to be found, which was discovered in 1932 and then 'lost' until 1973. Apollo asteroids are so small and faint that they are difficult to see except when close to Earth. They take more than a year to go round the Sun. Apollo itself is about 2 km/1.2 mi across. Apollo asteroids may collide with the Earth from time to time.

- In December 1994 an Apollo asteroid called 1994 XM1 passed 100,000 km/60,000 mi from Earth, the closest observed approach of any asteroid.

- A collision with an Apollo asteroid 65 million years ago has been suggested as one of the causes of the extinction of the dinosaurs.

See also: *Amor asteroids.*

Apollo project

United States project to land a person on the Moon. It achieved its goal on 20 July 1969, when Neil Armstrong was the first to set foot there. He was accompanied on the Moon's surface by 'Buzz' Aldrin; Michael Collins remained in the orbiting command module.

The *Apollo* programme was announced in 1961 by President John F Kennedy. The world's most powerful launcher, the *Saturn V* **rocket**, was built to launch the *Apollo* spacecraft, which carried three astronauts. When the spacecraft was in orbit around the Moon, two astronauts would descend to the surface in a lunar module to take samples of rock and set up experiments that would send data back to Earth. After three preparatory flights, *Apollo 11* made the first lunar landing. Five more crewed landings followed, the last in 1972. The total cost of the programme was over $24 billion.

Apollo missions

Mission	Launch date	Notes
1	—	During a preliminary check on the ground on 27 January 1967 the three crew were killed by a fire.
4	9 November 1967	Launched into an orbit around the Earth; the first time the *Saturn V* rocket was used.
7	11 October 1968	The first *Apollo* mission carrying a crew, *Apollo 7* was a test flight sent into orbit around the Earth.
8	21 December 1968	The first rocket to take a crew round the Moon.
9	3 March 1969	The lunar module was tested in orbit around the Earth.
10	18 May 1969	Lunar module successfully tested 14.5 km/9 mi above the surface of the Moon.

Apollo missions (*continued*)

Mission	Launch date	Notes
11	16 July 1969	Lunar module *Eagle* landed in the Sea of Tranquillity on 20 July 1969. The module remained on the Moon for 22 hours during which time the astronauts collected rocks, set up experiments, and mounted a US flag. After return the astronauts were quarantined in case they had brought unknown micro-organisms from the Moon. (None were ever detected.)
12	14 November 1969	Although the launcher was twice struck by lightning during takeoff, the lunar module made the second Moon landing.
13	11 April 1970	On the third day of the mission an explosion in one of the oxygen tanks cut off supplies of power and oxygen to the command module. The planned landing was abandoned, but the rocket coasted round the Moon before heading back to Earth. The crew splashed down safely on 17 April.
15	26 July 1971	First surface vehicle, the *Lunar Rover*, used on the Moon
17	7 December 1972	The last of the *Apollo* Moon landings. Detailed geological studies were carried out and large amounts of rock and soil were brought back.

Aquarius

A **zodiacal constellation** a little south of the celestial equator near **Pegasus**. The Sun passes through Aquarius from late February to early March.

- Aquarius means 'water-carrier', and is represented as a person pouring water from a jar.

- In astrology, the dates for Aquarius, the 11th sign of the zodiac, are between about 20 January and 18 February.

   See also: *precession.*

Aquila

A **constellation** on the celestial equator (see: **celestial sphere**). Its brightest star is first-magnitude **Altair**, flanked by the stars Beta and Gamma Aquiline. *Aquila* means 'eagle' in Latin. Nova Aquiline, which appeared in June 1918, was a **nova** that for a few days shone nearly as brightly as **Sirius**.

- In Greek mythology Aquila was the eagle of Zeus, king of the gods, and, according to some, the bird that tormented Prometheus for stealing fire from heaven.

- The Roman emperor Hadrian (AD 76–138) ordered that its southern stars should be formed into a separate constellation named in honour of Antinous, a young man who had sacrificed his life for the emperor.

- The German astronomer Johann Bayer (1572–1625) showed the new constellation with Aquila but called it **Ganymedes**.
See also: *celestial sphere.*

Arecibo Observatory

A **radio-astronomy observatory** located in Puerto Rico and operated by the US National Astronomy and Ionosphere Center. It is the home of the largest **radio telescope** in the world, which consists of a large metal dish housed in a natural hollow. The telescope began operations in 1963 and received major upgrades in 1974 and again in the mid-1990s.

DISH DETAILS

- The 305-m/1,000-ft diameter spherical reflector of the Arecibo dish is made up of nearly 40,000 perforated aluminium panels.
- Each panel can be adjusted to maintain a precise spherical shape that varies less than 3 mm/0.12 in over the entire surface.
- The dish reflects radio waves to an overhead receiver slung by cables from three towers. As the Earth turns during the day, the telescope scans a band of the sky. Moving the receiver makes the band wider, but the telescope cannot observe the whole sky.

Ariane

A **rocket** launch vehicle built by the **European Space Agency**. The launch site is at **Kourou** in French Guiana. *Ariane* is a three-stage rocket using liquid fuels. Small solid-fuel and liquid-fuel boosters can be attached to its first stage to increase carrying power. The first flight was in 1979. In October 1998 *Ariane 5* completed its first mission, launching a dummy **satellite** into **orbit**. Since 1984 *Ariane* has been operated commercially by Arianespace, a private company financed by European banks and aerospace industries.

Ariane 5, the latest and most powerful version of the launcher, was launched on 4 June 1996. It was intended to carry astronauts aboard the Hermes spaceplane. However, it went off course immediately after takeoff, turned on its side, broke into two and disintegrated. A fault in the software controlling the takeoff trajectory was to blame. A largely successful test flight for *Ariane 5* was accomplished in November 1997.

Aries

One of the **constellations** of the **zodiac**. It lies in the northern hemisphere between **Pisces** and **Taurus**, near **Auriga**. Its most distinctive feature is a curve of three stars of decreasing brightness. The brightest of these is Hamal, or Alpha Arietis, 65 **light years** from Earth. The Sun passes through Aries from late April to mid-May. The spring **equinox** once lay in Aries, but has now moved into Pisces through the effect of the Earth's **precession** (wobble).

- Aries is said to represent the winged ram whose golden fleece, in ancient Greek legend, was sought by Jason and the Argonauts.

- In astrology, the dates for Aries, the first sign of the zodiac, are between about 21 March and 19 April.

Aristarchus of Samos (c. 320 BC–c. 250 BC)

A Greek astronomer who was the first to argue that the Earth moves around the Sun. He was also the first astronomer to estimate the sizes of the Sun and Moon and their distances from the Earth. Aristarchus was born on the island of Samos and may have studied in Alexandria, where he died. His only surviving work is called *Magnitudes and Distances of the Sun and Moon*. He devised methods for finding the relative distances of the Sun and Moon that were geometrically correct but impractical because in his day the necessary observations could not be made sufficiently accurately.

Aristarchus believed that the Sun and the fixed stars are stationary and the **planets** – including the Earth – travel in circular **orbits** around the Sun and the apparent daily rotation of the sky is due to the Earth's rotation on its axis.

Aristarchus also believed that there is apparently no stellar **parallax** (shift in position of the stars in relation to each other) throughout the year because even the diameter of the Earth's orbit is insignificant in relation to the vastness of the universe.

asteroid or minor planet

Any one of many thousands of small bodies, composed of rock and iron, that orbit the Sun. Most lie in a belt between the orbits of **Mars** and **Jupiter**, and are thought to be fragments left over from the formation of the **Solar System**. About 100,000 may exist, but their total mass is only a few hundredths of the mass of the Moon. The asteroids include **Ceres** (the largest asteroid, 940 km/584 mi in diameter), **Vesta** (which has a light-coloured surface, and is the brightest as seen from Earth), **Eros**, and **Icarus**.

- Some asteroids are in orbits that bring them close to Earth, and some, such as the **Apollo asteroids**, even cross Earth's orbit; at least some of these may be remnants of former comets.
- One group, the **Trojan asteroids**, moves along the same orbit as **Jupiter**.
- One unusual asteroid, Chiron, orbits beyond **Saturn**.
- The Near Earth Asteroid Tracking (NEAT) system had detected more than 28,000 asteroids by August 1999.

Largest asteroids

Name	Diameter (km/mi)	Average distance from Sun (Earth = 1)	Orbital period (years)
Ceres	940/584	2.77	4.6
Pallas	588/365	2.77	4.6
Vesta	576/358	2.36	3.6
Hygieia	430/267	3.13	5.5
Interamnia	338/210	3.06	5.4
Davida	324/201	3.18	5.7

astrolabe

An astronomical instrument once used by sailors and astronomers. It is reputed to have been invented about 200 BC by the Greek astronomer **Hipparchus**. Some of the most beautiful examples were later constructed by Arab scientists. The astrolabe consisted of an engraved disc on which degree markings, lines, and circles were engraved. A metal mesh with points representing the brightest stars rotated over this, representing the daily movement of the sky. A rotating sighting arm could be used to measure the height of a star above the horizon.

astronomical unit (AU)

An astronomical unit of distance, symbol AU. It was originally defined as equal to the mean distance of the Earth from the Sun, but its value is now fixed by definition as 149,597,870 km/92,955,807 mi. It is useful for expressing the distance of planets from the Sun. Light travels 1 AU in approximately 8.3 minutes.

astronomy

The science of the celestial bodies: the Sun, the Moon, and the **planets**; the stars and **galaxies**; and all other objects in the universe. It is concerned with their positions, motions, distances, and physical conditions and with their origins and evolution. Astronomy is divided into fields such as **astrophysics**, celestial mechanics, and cosmology.

The first astronomers to develop scientific theories were the Greeks. Their ideas were kept alive during the European Dark Ages by Arab scholars, who themselves made detailed observations of the sky. The dawn of a new era came in 1543, when a Polish astronomer, Nicolaus **Copernicus**, published a work in which he argued that the description of planetary movements could be simplified by taking the Sun, not the Earth, to be the centre of the **Solar System**. The **telescope** came into use about 1608, and in the hands of an Italian mathematician and astronomer, **Galileo** Galilei, revolutionized astronomy. By the end of the 17th century, English physicist and mathematician Isaac **Newton** had established celestial mechanics, the science of the movement of celestial bodies under the influence of **gravity**.

In the 18th and 19th centuries astronomers discovered new planets, measured the distances to some of the nearer stars, and began to learn what celestial bodies are made of by means of **spectroscopy**, the analysis of light into its constituent wavelengths, or colours. By 1900 **astrophotography** had started to play a leading role in research. The 20th century saw the telescopic exploration of the galaxies beyond our own and the discovery that the universe is expanding from the **Big Bang**, a stupendous explosion about 15 billion years ago. Today astronomical observatories are placed on mountain-tops and in space, and robot spacecraft have visited all the planets except **Pluto**.

See also: *black hole, Fraunhofer lines, gamma-ray astronomy, Hubble Space Telescope, infrared astronomy, radio astronomy, ultraviolet astronomy, X-ray astronomy.*

astrophotography
Photography has been a vital part of the study of astronomy for almost 200 years. Before the development of photography, observations were gathered in the form of sketches made at the **telescope**. Modern-day electronic innovations, notably **CCDs** (charge-coupled devices), provide a more efficient light-gathering capability than photographic film as well as enabling information to be transferred to a computer for analysis.

- The first successful photograph of a celestial object was the daguerreotype plate of the Moon taken by an American, John W Draper (1811–1882), in March 1840.
- Several successful daguerrotypes were obtained prior to the introduction of wet-plate collodion about 1850.
- Dry plates were introduced in the 1870s, and in 1880 Henry Draper (1837–1882) obtained a photograph of the **Orion nebula**.
- The first successful image of a **comet** was obtained in 1882 by the Scottish astronomer David Gill (1843–1914), his plate also displaying excellent star images.

- Following this, Gill and J C Kapteyn (1851–1922) compiled the first photographic atlas of the southern sky, cataloguing almost half a million stars.

astrophysics

The study of the physics of stars, **galaxies**, **nebulae** and other celestial objects is known as astrophysics. It began with the development of **spectroscopy** in the 19th century, which allowed astronomers to analyse the composition of stars from their light. Astrophysicists view the universe as a vast natural laboratory in which they can study matter under conditions of temperature, pressure, and density that are difficult to attain on Earth.

atmosphere

A mixture of gases surrounding a **planet**. Planetary atmospheres are prevented from escaping by the pull of **gravity**. See entries for individual planets for descriptions of their atmospheres. The lowest layer of the Earth's atmosphere consists of nitrogen (78%) and oxygen (21%), both in molecular form (that is, they consist of molecules, each made up of two atoms bonded together) and 1% argon. Small quantities of other gases are important to the chemistry and physics of the Earth's atmosphere; these include water and carbon dioxide. The atmosphere plays a major part in the various cycles of nature (water cycle, carbon cycle, and nitrogen cycle). Other atmospheric ingredients are found in particular localities: gaseous compounds of sulphur and nitrogen in towns, salt over the oceans; and everywhere dust composed of inorganic particles, decaying organic matter, tiny seeds and pollen from plants, and bacteria. Of particular importance are chlorofluorocarbons (CFCs) that damage the **ozone layer**.

Atmospheric pressure decreases continuously with altitude. However, the change in temperature with altitude is more complex, and is the basis for dividing the atmosphere into four layers: the **troposphere**, **stratosphere**, **mesosphere**, and **thermosphere**.

SWOLLEN ATMOSPHERE

During periods of intense solar activity, the atmosphere swells outwards; there is a 10–20% variation in atmosphere density. One result is to increase drag on satellites, making it impossible to predict exactly the time of their re-entry into the atmosphere.

See also: *exosphere, ionosphere, magnetosphere.*

Auriga

A **constellation** of the northern hemisphere, represented as a charioteer. Its brightest star is the first-**magnitude Capella**, about 45 **light years** from Earth; Epsilon Aurigae is an eclipsing **binary star** with a **period** of 27 years, the longest of its kind (last eclipse 1983).

The charioteer is usually represented as a man holding a bridle in his right hand and supporting a goat and kids on his left arm. He is said to be Erichthonius, a legendary king of Athens who invented the four-horse chariot.

aurora

Glowing coloured lights in the night sky near the Earth's magnetic poles, occurring at heights of over 100 km/60 mi. Although aurorae are usually restricted to the polar skies, fluctuations in the **solar wind** occasionally cause them to be visible at lower **latitudes**. An aurora is usually in the form of a luminous arch with its apex towards the magnetic pole followed by arcs, bands, rays, curtains, and **coronas**, usually green but often showing shades of blue and red, and sometimes yellow or white.

- The aurora is called aurora borealis ('northern lights') in the northern hemisphere and aurora australis ('southern lights') in the southern hemisphere.

- Aurorae are caused by fast streams of charged particles from **solar flares** and other events on the Sun. These are guided by the Earth's magnetic field towards the north and south magnetic poles, where they enter the upper atmosphere and bombard the gases in the atmosphere, causing them to emit visible light.

See also: *magnetic field, magnetic storm, magnetosphere.*

Australia Telescope

A giant system of linked **radio telescopes** in New South Wales, Australia, consisting of six 22-m/72-ft antennae at Culgoora, a similar antenna at Siding Spring Mountain, and the 64-m/210-ft Parkes radio telescope. Together they have the resolving power (ability to see detail) of a single dish 300 km/186 mi across. The system was set up in 1993 and is operated by the Commonwealth Scientific and Industrial Research Organization (CSIRO).

azimuth

The direction of an object, on the ground or in the sky, in relation to north. It is defined as the angular distance eastwards along the horizon, measured from due north, to the point on the horizon directly beneath the object.

See also: *altitude, celestial sphere.*

Baikonur

A launch site for spacecraft, located at Tyuratam, Kazakhstan, near the Aral Sea: the first **satellites** and all Soviet space probes and crewed *Soyuz* missions were launched from here. It covers an area of 12,200 sq km/4,710 sq mi, making it much larger than its US equivalent, the Kennedy Space Center at **Cape Canaveral**.

Baily's beads

Bright spots of sunlight seen around the edge of the Moon for a few seconds immediately before and after totality in a **total eclipse** of the Sun, caused by sunlight shining between mountains at the Moon's edge. Sometimes one bead is much brighter than the others, producing the so-called diamond-ring effect. The effect was described in 1836 by the English astronomer Francis Baily (1774–1844), a wealthy stockbroker who retired in 1825 to devote himself to astronomy.

Balmer series

Five lines (wavelengths) in the visible part of the hydrogen spectrum. They are named after Johann Jakob Balmer (1825–1898), a Swiss physicist and mathematician who in 1885 published an equation that described the four visible spectral lines of hydrogen that were then known.

Balmer used his equation to predict unknown spectral lines. He said there was a fifth line at the limit of the visible spectrum, which was soon detected and measured. And he further predicted the existence of other hydrogen spectral lines beyond the visible spectrum. His simple formula played a central role in the development of **spectroscopy** and atomic theory.

Barnard's star

The second-closest star to the Sun, six **light years** away in the **constellation Ophiuchus**. It is a faint red dwarf of 10th **magnitude**, visible only through a **telescope**. It is named after the US astronomer Edward E Barnard (1857–1923), who discovered in 1916 that it has the fastest **proper motion** of any star, crossing 1° of sky every 350 years. Some observations suggest that Barnard's star may be accompanied by **planets**.

barred spiral galaxy

A **spiral galaxy** that has a straight bar of stars across its centre, from the ends of which the spiral arms emerge. As in other spiral galaxies, the arms contain gas and **dust** from which new stars are still forming. About half of all spirals are barred spirals.

basalt

The commonest volcanic igneous rock in the **Solar System** – that is, rock formed from cooling magma or lava, and solidifying from a molten state. Much of the surfaces of the terrestrial planets **Mercury**, **Venus**, Earth, and **Mars**, as well as the Moon, are composed of basalt.

- Earth's ocean floor is virtually entirely made of basalt. It is usually dark grey but can also be green, brown, or black. Its essential constituent minerals are calcium-rich feldspar and calcium and magnesium-rich pyroxene.

- The dark-coloured lowland **mare** regions of the Moon are underlain by basalt. Lunar mare basalts have higher concentrations of titanium and zirconium and lower concentrations of volatile elements such as potassium and sodium relative to terrestrial basalts.

- Martian basalts have low ratios of iron to manganese relative to terrestrial basalts, as judged from some Martian **meteorites** and spacecraft analyses of rocks and soils on the Martian surface.

Bellatrix or Gamma Orionis

A star of the second **magnitude** that marks the western shoulder of **Orion**.

Bell-Burnell, (Susan) Jocelyn (1943–)

British radio astronomer who in 1967 discovered the first **pulsar** (rapidly flashing star), while part of a team led by Antony Hewish at the **Mullard Radio Astronomy Observatory**, Cambridge. Jocelyn Bell-Burnell was born in Belfast, Northern Ireland, near the Armagh Observatory, where she spent much time as a child. Bell-Burnell spent her first two years as a research student at Cambridge building a **radio telescope** that was specially designed to track **quasars**. The telescope had the ability to record rapid variations in signals. In 1967 she noticed an unusual signal, which turned out to consist of a rapid set of pulses that occurred precisely every 1.337 sec. Within a few months she located three other similar sources in widely spaced locations in our **Galaxy**. It seemed that the signals were being emitted by a special kind of object, which was given the name 'pulsar', from 'pulsating star'.

Betelgeuse or Alpha Orionis

Red **supergiant** star in the constellation of **Orion**. It is the tenth-brightest star in the night sky, although its brightness varies. It is 1,100 million km/ 700 million mi across, about 800 times as wide as the Sun. It is over 10,000 times as luminous as the Sun, and lies 650 **light years** from the Sun. Light takes 60 minutes to travel across the giant star. It was the first star whose angular diameter was measured with the **Mount Wilson** interferometer in 1920.

Big Bang

The hypothetical explosion that marked the origin of the universe as we know it. At the earliest moment that present-day physics can speak of with confidence, when the universe was a split second old, the entire universe of matter and energy that we can observe today was squeezed into the size of a pea. This hot, superdense matter and energy rapidly expanded and cooled, producing the expanding universe of today. The current rate of expansion of the universe suggests that the Big Bang took place 10–20 billion years ago. According to a modified version of the Big Bang, called the inflationary theory, the universe briefly underwent an especially rapid period of expansion within a tiny fraction of a second of the Big Bang, which accounts for its current remarkable uniformity.

See also: *red shift, Planck time.*

binary star

A pair of stars moving in **orbit** around their common centre of mass. Observations show that most stars are binary, or even multiple – for example, the nearest star system to the Sun, **Alpha Centauri**. Alpha Centauri consists of a star almost identical to the Sun accompanied by another one about a third as bright, and closer to it than **Neptune** is to the Sun. Each of these stars appears to describe an ellipse about the other in about 80 years. A third, much fainter star, **Proxima Centauri**, is too far away to disturb their mutual orbit appreciably. There are several kinds of binary star, which differ in the way they are detected by astronomers:

- *Visual binaries* can be seen by direct telescopic observation.

- *Eclipsing binaries* are discovered by changes in brightness caused by each member of the system passing in front of the other.

- *Interferometric binaries* are detected by **interferometry**.

- *Astrometric binaries* are detected by periodic variations in **proper motion** (that is, 'sideways movement').

- *Spectroscopic binaries* are detected when periodic changes in radial velocity (movement towards or away from the Earth) show up in their spectra.

See also: *spectroscopy.*

binoculars

An optical instrument consisting of two **telescopes**, with which an object can be viewed with both eyes. Field glasses and opera glasses are examples. Each telescope contains lenses and **prisms** to 'fold' the light beams so that the instrument can produce a large magnification while being compact. The wide separation of the main, or object, lenses enhances the stereoscopic effect in the image as well as magnifying it. The first binocular telescope was constructed in 1608 by Hans Lippershey, the Dutch inventor reputed to have made the first telescope.

black dwarf

A star that has cooled from the **white dwarf** state so that it is now too dim to be seen.

black hole

An object in space whose **gravity** is so intense that nothing can escape from it, not even light. Some black holes are thought to form when massive stars collapse at the end of their lives. A black hole sucks in more matter, including other stars, from the space around it. Matter that falls into a black hole is squeezed to enormous density at the centre of the hole. Black holes can be detected because gas falling towards them becomes so hot that it emits **X-rays**.

Black holes containing the mass of millions of stars are thought to lie at the centres of **quasars**. **Satellites** have detected X-rays from a number of objects that may be black holes, but only a small number of likely black holes have been identified in our **Galaxy**.

Microscopic black holes may have been formed in the extreme pressures and temperatures of the **Big Bang**. The British physicist Stephen **Hawking** has shown that such tiny black holes could 'evap-

black hole *An artist's interpretation of a black hole.*

orate' and explode in a flash of energy. Some possible locations of black holes:

- Nova Muscae is a **nova** or 'new star', identified as a black hole in 1992. It lies approximately 18,000 **light years** from Earth.

- V404 Cygni, is a 'recurrent' (repeating) nova, a possible black hole discovered in 1992.

- A0620-00, an X-ray source in **Monoceros**, brightened by 100,000 times in 1975. It is one of the best black-hole candidates in the Galaxy.

- In 1997 the **Hubble Space Telescope** discovered evidence of a black hole of 300 million times the mass of the Sun. It is located in the middle of galaxy M84, about 50 million light years from Earth.

    See also: *supernova.*

blink comparator
A device that helps astronomers to detect changes in celestial objects. The comparator works with two photographs of the same part of the sky taken weeks, months, or even years apart. It 'blinks' the pictures by showing them alternately to the viewer in rapid succession. A star that has altered in brightness will seem to flash; an **asteroid** that has moved will seem to jump back and forth between its two positions.

blue-shift
An alteration in the **spectrum** of an object when it is moving at high speed towards the observer, or the observer is moving towards it. All wavelengths appear shortened, so that visible wavelengths move towards the blue end of the spectrum, while blue wavelengths move into the ultraviolet. Some infrared radiation is shortened enough to become visible. Some stars show blue-shifts, and some relatively nearby galaxies. The great majority of galaxies show **red-shifts**.

Bode's law or Titius–Bode law
A numerical sequence that gives the approximate distances of the **planets** from the Sun, in **astronomical units** (distance between Earth and Sun = one astronomical unit). The series begins arbitrarily with 0 and 3. More terms are found by doubling:

> 0, 3, 6, 12, 24, ...

and adding 4 to each term to give:

> 4, 7, 10, 16, 28, ...

and then dividing by 10:

> 0.4, 0.7, 1.0, 1.6, 2.8, ...

The sequence fits pretty well, as far as **Uranus**. There is a gap at 2.8, between **Mars** and **Jupiter**, which was filled when the first **asteroid** was discovered. The 'law' breaks down for **Neptune** and **Pluto**, and is not now regarded as having any real significance. The relationship was first noted in 1772 by the German mathematician Johann Titius (1729–1796) and became better known when it was discussed by the German astronomer Johann Bode (1747–1826).

Application of Bode's Law

	Mercury	Venus	Earth	Mars	Asteroids
Bode's Law	0.4	0.7	1.0	1.6	2.8
Actual distance (Earth = 1)	0.4	0.7	1.0	1.5	-

	Jupiter	Saturn	Uranus	Neptune	Pluto
Bode's Law	5.2	10	19.6	38.8	77.2
Actual distance (Earth = 1)	5.2	9.5	19.2	30.1	39.4

Bok globule

A small, spherical object in a **nebula**, with a mass comparable to that of the Sun. Bok globules are possibly gas clouds in the process of condensing into stars. They were first discovered by the Dutch astrophysicist Bart Bok (1906–1983).

bolide

A very bright **meteor**, which sometimes breaks up into fragments.

Boötes

A **constellation** of the northern sky. Its brightest star is Arcturus (or Alpha Boötes), which is about 37 **light years** from Earth. Epsilon Boötes, a double star with blue and yellow components, was called Pulcherrima (most beautiful) by F G W Struve and was used by William **Herschel** in an unsuccessful attempt to detect a **parallax** displacement and thus determine a stellar distance.

Boötes is represented by a herdsman driving a bear (Ursa Major) around the pole. The herdsman is assisted by the neighbouring Canes Venatici, 'the hunting dogs'.

Brahe, Tycho (1546–1601)

Danish astronomer whose accurate observations of the **planets** enabled the German astronomer and mathematician Johannes **Kepler** to prove that planets **orbit** the Sun in **ellipses**. Brahe observed the 'new star' that blazed forth in **Cassiopeia** from November 1572. It was a **supernova**, which was bright enough to be seen by day, and was visible for over a year. Brahe gave an account of the star in *De Nova Stella* (1573), in which he pointed out that his

Brahe wore a silver nose after his own was cut off in a duel, and took an interest in alchemy. In 1576 Frederick II of Denmark gave him the island of Hven, where he set up an observatory called Uraniborg. He was visited there by many notable persons, including James VI of Scotland, later James I of England, who wrote a poem in his honour. Brahe left Denmark in 1597 after he had failed to retain the favour of King Frederick's successors. He found a new patron in the Holy Roman Emperor Rudolf II.

Brahe *Danish astronomer Tycho Brahe (1546–1601).*

observations showed it to be farther away than the Moon, and thus in those realms where, according to Aristotelian philosophy, no change could take place. He observed the bright **comet** of 1577, and came to the conclusion that the comet's orbit must be elongated, which conflicted with the belief that the celestial bodies move in spheres. Brahe, the last great astronomer to reject the **heliocentric theory** of **Copernicus**, tried to compromise, suggesting that all the **planets** revolved around the Sun, while the Earth remained stationary.

bridges between galaxies
Streams of gas and stars linking neighbouring **galaxies**. They are drawn out of the galaxies when they pass close to each other or even collide. An example is a bridge of hydrogen gas called the Magellanic Stream, which flows from the **Magellanic Clouds** towards our **Galaxy**.

brown dwarf
A faintly glowing gaseous object, less massive than a star, but heavier than a **planet**. Brown dwarfs are 'failed stars', formed without enough mass to ignite nuclear reactions at their centres. They shine by heat released during their contraction from the gas cloud in which they were born. Some astronomers believe that vast numbers of brown dwarfs exist throughout the **Galaxy**. Because of the difficulty of detection, none were spotted until 1995, when US astronomers discovered a brown dwarf, GI229B, in the **constellation** Lepus. It is about 20–40 times as massive as **Jupiter** and emits only 1% of the radiation of the smallest and faintest known star. In 1996 UK astronomers discovered four possible brown dwarfs within 150 light years of the Sun, and more are being discovered all the time.

C

caldera

A very large basin-shaped crater. Calderas are found at the tops of **volcanoes**, where the original peak has collapsed into an empty chamber beneath. The basin, many times larger than the original volcanic vent, may be flooded, producing a crater lake, or the flat floor may contain a number of small volcanic cones, produced by volcanic activity after the collapse. Typical calderas are Kilauea, Hawaii; Crater Lake, Oregon, USA; and the summit of Olympus Mons, on **Mars**. Calderas should be distinguished from **craters**.

Callisto

The second-largest **satellite** of **Jupiter**. Its surface is covered with large **craters**. In 1997 the space probe *Galileo* detected molecules containing both carbon and nitrogen atoms on the surface of Callisto. Their presence may indicate that Callisto harboured life at some time.

CALLISTO: STATISTICS

Diameter	Distance from centre of planet	Period
4,800 km/3,000 mi	1.9 million km/1.2 million mi	16.7 days

Canals of Mars

A network of straight lines apparently criss-crossing the surface of **Mars**, 'discovered' by the US astronomer Percival Lowell (1855–1916). He claimed these must be a network of artificial channels leading water from the poles to the arid deserts, dug by a civilization battling against extinction. He even believed he could see the lines growing thicker or thinner in different seasons as vegetation bordering the canals grew or died off. Lowell's vision of an inhabited Mars spawned a genre of science-fiction stories, most notably *The War of the Worlds* by H G Wells (1866–1946). Other astronomers, though, failed to see the canals and observations from *Mariner* and *Viking* spacecraft have shown that they do not exist. They were an illusion caused by a naked-eye observer straining to make out tiny details at the very limits of observability.

Cancer
The faintest of the **zodiacal constellations** (its brightest stars are of the fourth **magnitude**). It lies in the northern hemisphere between **Leo** and **Gemini**, and is represented as a crab. The Sun passes through the constellation during late July and early August.

- Cancer's most distinctive feature is the **open cluster** Praesepe, popularly known as the Beehive, visible to the naked eye as a nebulous patch.

- In Chaldaean and Platonist philosophy Cancer was the 'gate of men', through which souls descended into human bodies, eventually returning to heaven through **Capricornus**, the 'gate of the gods'.

- In astrology, the dates for Cancer are between about 22 June and 22 July.

Canis Major
A brilliant **constellation** of the southern hemisphere, represented as one of the two dogs following at the heel of **Orion**. Its main star, **Sirius**, is the brightest star in the night sky. Epsilon Canis Majoris is also of the first **magnitude**, and there are three second-magnitude stars.
See also: *Canis Minor.*

Canis Minor
A small **constellation** along the celestial equator (see **celestial sphere**), represented as the smaller of the two dogs of **Orion** (the other dog being **Canis Major**). Its brightest star is the first-**magnitude Procyon**.

Procyon and Beta Canis Minoris form what Arab astronomers called 'the short cubit', in contrast to 'the long cubit' formed by Castor and Pollux (Alpha and Beta Geminorum).

Canopus or Alpha Carinae
The second-brightest star in the night sky (after **Sirius**), lying in the southern constellation **Carina**. It is a yellow–white **supergiant** about 120 **light years** from the Sun, and thousands of times more luminous. It may have been named in honour of the chief pilot of the Greek fleet that sailed against Troy, or the name may be derived from two Coptic words signifying 'golden earth'.

Cape Canaveral
A promontory on the Atlantic coast of Florida, USA, 367 km/228 mi north of Miami, used as a **rocket** launch site by NASA. It was known as Cape Kennedy from 1963 to 73. The Kennedy Space Center is nearby.

Capella or Alpha Aurigae

The brightest star in the **constellation Auriga** and the sixth- brightest star in the night sky. It is a visual and spectroscopic **binary** that consists of a pair of yellow **giant stars** 45 **light years** from the Sun, orbiting each other every 104 days.

'Capella' is a Latin word meaning 'the little nanny goat'. The nanny-goat's kids are the three adjacent stars Epsilon, Eta, and Zeta Aurigae.

Capricornus

The **zodiacal constellation** in the southern hemisphere following (that is, east of) **Sagittarius**. It is represented as a fish-tailed goat, and its brightest stars are of the third **magnitude**. The Sun passes through it from late January to mid-February.

- In Chaldaean and Platonist philosophy Capricornus was regarded as the 'gate of the gods' through which souls ascended to heaven, having descended through Cancer.

- In astrology, the dates for Capricornus (popularly known as Capricorn) are between about 22 December and 19 January.

carbon–nitrogen cycle

A sequence of nuclear reactions in stars that releases energy and makes them shine. In the cycle, carbon atoms act as a catalyst to convert four **hydrogen** atoms into one **helium** atom with the release of energy. The carbon– nitrogen cycle is the dominant energy source for ordinary stars of mass greater than about 1.5 times that of the Sun. Cooler stars, including the Sun, get most of their energy from the **proton–proton cycle**.

Cassegrain telescope or Cassegrain reflector

A type of reflecting **telescope** in which **light** collected by a concave primary mirror is reflected on to a convex secondary mirror, which in turn directs it back through a hole in the primary mirror to a focus behind it. As a result, the telescope tube can be kept short, and equipment for analysing and recording starlight can be mounted behind the main mirror. All modern large astronomical telescopes are of the Cassegrain type. It is named after a French astronomer, Cassegrain (c. 1650–c. 1700), who first devised it as an improvement to the simpler Newtonian telescope. (*See illustration on p. 26.*)

Cassini, Giovanni Domenico (1625–1712)

Italian-born French astronomer who made many important observations. He discovered four **satellites** of **Saturn** and the gap in the rings of Saturn

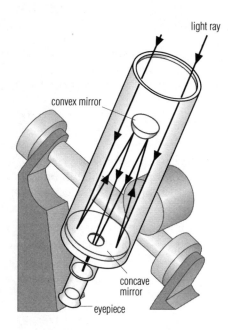

light ray

convex mirror

concave mirror

eyepiece

Cassegrain telescope *In this reflecting telescope, a hole in the centre of the concave main mirror allows light reflected by the convex secondary mirror to reach the eyepiece (or a camera).*

that is now called the **Cassini division**. Cassini correctly suggested that the rings of Saturn were composed of myriads of tiny satellites. He refused to accept the **Copernican system** and rejected the concept of a finite speed of light, although the speed of light was estimated by a Danish astronomer, Ole Römer (1644–1710), using Cassini's own data. Cassini made many observations of details on the lunar surface.

See also: *planetary rings.*

Cassini division

A prominent gap in the rings of **Saturn**. It was discovered by the Italian-born French astronomer Giovanni **Cassini** in 1675. Some 5,000 km/3,100 mi wide, the Cassini division separates the faint outer A-ring from the brighter inner B-ring. It is now known that the division is

CASSINI'S LIFE

1625 Born near Nice (then in Italy).

1650 He is made professor of astronomy at the University of Bologna, at the age of 25.

1669 He departs for France at the invitation of King Louis XIV, to construct and run the Paris Observatory.

1672 Cassini takes advantage of a good opposition of **Mars** to determine the distance between the Earth and that **planet**. From the result, Cassini is able to deduce many other astronomical distances.

1710 Cassini goes blind, and his son Jacques Cassini (1677–1756) succeeds him at the Paris Observatory. Later his grandson, César Cassini (1714–1784) and great-grandson, Jacques Dominique (1748–1845) serve as directors of the Observatory.

1712 Cassini dies.

caused by the gravitational influence of Mimas, one of the **satellites** of Saturn.

See also: *planetary rings.*

Cassini–Huygens **mission**
Unmanned space mission by the US agency **NASA** and the **European Space Agency** (ESA) to the planet **Saturn**. The craft was launched in October 1997 on a *Titan 4* rocket. Electricity to power the craft during the mission is supplied by 32 kg/ 70 lb of plutonium. This was the largest amount of radioactive material ever to be sent into space, and provoked fears of contamination should *Cassini*, or its **rocket**, malfunction. The spacecraft will go into orbit around Saturn in 2004. There it will drop off a sub-probe, *Huygens*, to land on Saturn's largest moon, **Titan.**

Cassiopeia
A prominent **constellation** of the northern hemisphere. It has a distinctive W-shape, and contains one of the most powerful radio sources in the sky, Cassiopeia A.

* Cassiopeia A is the remains of a **supernova** (star explosion) that occurred *c.* AD 1702, too far away to be seen from Earth.

* Tycho **Brahe** observed a new star in Cassiopeia in 1572; it was probably a supernova, since it was visible in daylight and outshone **Venus** for ten days.

* Cassiopeia is named after the mother of **Andromeda**, a princess in Greek mythology.

Castor or Alpha Geminorum
The second brightest star in the constellation **Gemini** and the 23rd-brightest star in the night sky. Along with the brighter **Pollux**, it forms a prominent pair at the eastern end of **Gemini**, representing the heads of the twins.

SEXTUPLE STAR SYSTEM
Second-magnitude Castor is 45 light years from Earth, and is one of the finest binary stars in the sky for small telescopes. The two main components orbit each other over a period of 467 years. A third, much fainter, star orbits the main pair over a period probably exceeding 10,000 years. Each of the three visible components is a spectroscopic binary, making Castor a sextuple star system.

CCD (charge-coupled device)

Device for forming images electronically. It is based on a semiconducting material, such as silicon, that releases **electrons** when struck by incoming light. The brighter the light is, the more electrons are given off. The number of electrons stored in each pixel (picture element) is measured electronically and read into a computer at the end of the exposure.

- CCDs have now almost entirely replaced photographic film for **astrophotography**, where extreme sensitivity to light is paramount.

- However, CCD images are expensive and very small in size compared to photographic plates. Photographic plates are better suited to wide-field images, whereas CCDs are used for individual objects, which may be very faint, within a narrow field of sky.

celestial sphere

The imaginary sphere surrounding the Earth, on which the celestial bodies are imagined to lie when specifying their positions. The positions of bodies such as stars, **planets**, and **galaxies** are specified by their coordinates on the celestial sphere. The celestial sphere appears to rotate once around the Earth each day, as a result of the rotation of the Earth on its axis.

- The celestial poles lie directly above the Earth's poles.

- The celestial equator lies over the Earth's equator.

- The equivalent of **latitude** on the celestial sphere is called **declination**

- The equivalent of **longitude** on the celestial sphere is called **right ascension**. It is measured in hours from 0 to 24.

(See illustration on opposite page.)

Centaurus

A large, bright **constellation** of the southern hemisphere, represented as a centaur. Alpha and Beta Centauri are both of the first **magnitude** and, like Alpha and Beta Ursae Majoris (see: **Ursa Major**), are known as the Pointers, as a line joining them leads to **Crux**.

- The brightest star in Centaurus, **Alpha Centauri**, is a triple star that includes the closest star to the Sun, **Proxima Centauri**, which is only 4.3 **light years** away.

- Omega Centauri, which is just visible to the naked eye as a hazy patch, is the largest and brightest **globular cluster** of stars in the sky, 16,000 light years away.

- Centaurus A, a galaxy 15 million light years away, is a strong source of radio waves and **X-rays.**

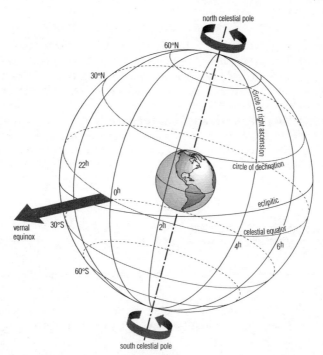

north celestial pole

60°N

30°N

circle of right ascension

22ʰ

circle of declination

0ʰ

ecliptic

vernal
equinox

30°S

2ʰ

celestial equator

4ʰ

6ʰ

60°S

south celestial pole

celestial sphere *The main features of the celestial sphere. The equivalents of latitude and longitude on the celestial sphere are declination and right ascension. Declination runs from 0° at the celestial equator to 90° at the celestial poles. Right ascension is measured in hours eastwards from the vernal equinox, one hour corresponding to 15° of longitude.*

Cepheid variable

Any one of a class of yellow **supergiant** stars that vary regularly in brightness every few days or weeks as a result of pulsations in size. The time that a Cepheid variable takes to pulsate is directly related to its average brightness; the longer the pulsation **period**, the brighter the star.

- The relationship between pulsation time and brightness is called the period–luminosity law. It was discovered by US astronomer Henrietta **Leavitt** and is one of the most important discoveries in astronomy. It allows astronomers to use Cepheid variables as 'standard candles' to measure distances in our **Galaxy** and to nearby **galaxies**.

- Cepheid variables are named after the star Delta Cephei, whose light variations were observed in 1784 by English astronomer John Goodricke (1764–1786).

Ceres
The largest **asteroid**, 940 km/584 mi in diameter, and the first to be discovered (by the Italian astronomer Giuseppe Piazzi (1746–1826) in 1801). Ceres orbits the Sun every 4.6 years at an average distance of 414 million km/257 million mi. Its mass is about one-seventieth of that of the Moon.

Cerro Tololo Inter-American Observatory
A major **observatory** on Cerro Tololo mountain in the Andes. It is operated by AURA, the Association of Universities for Research into Astronomy, and its main instrument is a 4-m/158-in reflector, a twin of the Mayall reflector at **Kitt Peak**, Arizona,USA. It began operation in 1974.

Cetus
A large **constellation** on the celestial equator (see **celestial sphere**), represented as a sea monster or a whale. Despite its size, Cetus contains few bright stars.

- Cetus is named after the sea monster sent to devour Andromeda in Greek mythology.
- The constellation contains the variable star **Mira**, which has a long **period**. Mira is sometimes the most conspicuous object in Cetus, but it is more usually invisible to the naked eye.
- The star Tau Ceti is one of the nearest stars, only 11.9 **light years** from the Earth.

Chandrasekhar limit or Chandrasekhar mass
The maximum possible mass of a **white dwarf** star. The limit depends slightly on the composition of the star but is approximately 1.4 times the mass of the Sun. A white dwarf heavier than the Chandrasekhar limit would collapse under its own weight to form a neutron star (see **pulsar**) or a **black hole**. The limit is named after the Indian-born US astrophysicist Subrahmanyan Chandrasekhar (1910–1995) who developed the theory of white dwarfs in the 1930s.

Charon
Satellite of **Pluto**, discovered in 1978 by James Walter Christy (1938–). It is about half the diameter of Pluto, making it the largest moon in relation to its parent **planet** in the **Solar System**. It completes one orbit in the same time that Pluto takes to spin on its axis. Charon is composed mainly of ice. Some astronomers have suggested that Pluto was a former moon of **Neptune** that escaped, but it is more likely that it was an independent body that was captured.

CHARON: STATISTICS

Diameter	Distance from centre of planet	Period
1,200 km/750 mi	20,000 km/12,500 mi	6.39 days

chondrite

A type of **meteorite** that contains **chondrules**, which are small spheres made up of the silicate minerals olivine and orthopyroxene.

chondrule

A small, round mass of silicate material, about 1 mm/0.04 in diameter, found in certain stony **meteorites**, which are called **chondrites**. Chondrules are thought to be mineral grains that condensed from hot gas in the early **Solar System**, most of which were later incorporated into larger bodies from which the **planets** formed.

chromatic aberration

A distortion of an optical image in which the image is surrounded by coloured fringes, because **light** of different colours is brought to different focal points by a lens. An achromatic lens is a combination of lenses made from materials of different refractive indexes, constructed in such a way as to minimize chromatic aberration. A reflecting **telescope** does not suffer from chromatic aberration.

See also: *refraction, spherical aberration.*

chromosphere

A layer of hot gas around the Sun. It consists mostly of **hydrogen** and is about 10,000 km/6,000 mi deep. Beneath it is the visible surface of the Sun (the **photosphere**). The chromosphere is pinkish red, and becomes visible as a red rim around the dark Moon during a **total eclipse** of the Sun. Its name comes from the Greek words meaning 'colour' and 'sphere'.

Clementine

US lunar probe launched on 25 January 1994. It spent two months surveying the Moon and discovered an enormous **crater** on the Moon's far side before being sent on to intercept **asteroid** Geographos. The mission was abandoned in June 1994 after a faulty thruster failed to shut off during orbital manoeuvres.

COBE or Cosmic Background Explorer

US astronomical satellite launched in 1989 to study **cosmic background radiation**. It confirmed that the radiation has a temperature of 2.735 K (–270.4°C/–454.7°F) and revealed ripples in the background radiation believed to mark the first stage in **galaxy** formation.

See also: *satellite, artificial.*

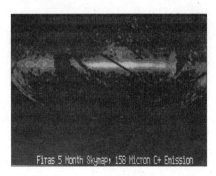

Firas 5 Month Skymap, 158 Micron C+ Emission

COBE *This image of the Milky Way was constructed from preliminary data obtained by the Far Infrared Absolute Spectrophotometer (FIRAS) on NASA's Cosmic Background Explorer (COBE).*

colour index

A measure of the colour of a star or other celestial object made by observing its apparent brightness through different coloured filters. It is defined as the difference between the **magnitudes** of the star as measured through two standard photometric filters. Colour index is directly related to the surface temperature of a star and its spectral classification.

coma

The hazy cloud of gas and dust that surrounds the nucleus of a **comet**. It is typically up to 100,000 km/60,000 mi wide.

Coma Berenices

A **constellation** of the northern hemisphere, represented as Queen Berenice's hair. Many of the brighter stars in the constellation belong to an extended **open cluster**, but the term 'Coma cluster' usually refers to a concentration of the **galaxies** that abound in this part of the sky.

The constellation was named in the 3rd century BC to appease Queen Berenice, after her hair, which she had sacrificed to Aphrodite, had been stolen from the temple.

comet

A small, icy body orbiting the Sun, usually on a highly elliptical path. It becomes visible when it approaches the Sun and gives off gas and dust, forming a **coma** (a 'halo') and one or more tails. A comet consists of a central nucleus a few kilometres across, and has been likened to a dirty snowball because it consists mostly of ice mixed with dust. As a comet approaches the Sun it is invisible until its nucleus heats up, releasing gas and dust and forming the coma. Gas and dust stream away from the coma to form one or more tails, which may extend for millions of kilometres.

Comets are believed to have been formed at the birth of the **Solar System**. Billions may exist in a halo (the **Oort cloud**) beyond **Pluto**. The gravitational effect of passing stars pushes a few of these towards the Sun, where they develop comas and tails. Most of these become long-period comets,

comet *Photograph of Halley's comet taken in 1986.*

with periods greater than 200 years. Their orbits are tilted at all angles to the **ecliptic**, and about equal numbers of them are direct and **retrograde**.

A few 'periodic' comets have their orbits altered by the gravitational pull of the **planets** so that they reappear every 200 years or less. They move mainly in direct orbits close to the plane of the Solar System and mostly have **aphelion** distances close to the orbit of **Jupiter**. There are similar, but

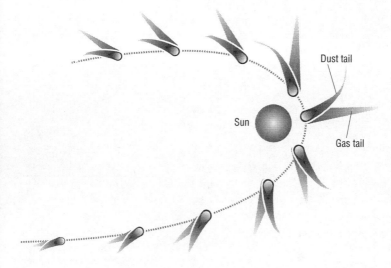

comet *A comet's tail always points away from the Sun, so a comet travels tail first if it is travelling away from the Sun. The further away from the Sun a comet travels the smaller its tail becomes.*

less numerous, families associated with the other major planets. There is little doubt that they are long-period comets that have been captured. Some come from the **Kuiper Belt**, a zone just beyond **Neptune**.

See also: *asteroid, meteoroid.*

COMET FACTS

- Of the 800 or so comets whose orbits have been calculated, about 160 are periodic.
- The comet with the shortest known period (time to go around the Sun once) is Encke's comet, which orbits the Sun every 3.3 years.
- A dozen or more comets are discovered every year, many by amateur astronomers.
- Most comets are named after their discoverers. When first discovered a comet is temporarily named with the year of its discovery and a letter indicating the order of discovery. Later the comet is numbered in order of its perihelion passage (the moment when it is nearest to the Sun); for example, Comet Arend–Roland was first called 1956h but became 1957 III.
- Some of the long-period comets are probably fragments of bigger comets that broke into pieces as they passed round the Sun. One of these groups consists of the bright Sun-grazing comets. These are comets that have passed through the solar corona and near perihelion were bright enough to be observed in daylight.
- Few of the short-period comets develop bright tails, because they are 'worn out' and have lost most of their volatile material.
- The brightest periodic comet, Hale–Bopp, flew past the Earth in March 1997. NASA launched rockets to study the comet, which came within 196 million km/122 million mi of the Earth. Its icy core is estimated to be 40 km/25 mi wide.
- In 1999 NASA approved the $240 million Deep Impact project to study comet structure. The aim of the project is to fire a projectile weighing 500 kg/1,100 lb into comet P/Tempel 1 in July 2005. The material hurled from the comet on impact will be analysed to determine the comet's makeup. The resulting crater will be approximately 120 m/396 ft in diameter and 25 m/82.5 ft deep.

Major comets

Name	First recorded sighting	Orbital period (years)	Interesting facts
Halley's Comet	240 BC	76	Parent of Eta Aquarid and Orionid meteor showers
Comet Tempel–Tuttle	1366	33	Parent of Leonid meteors
Biela's Comet	1772	6.6	Broke in half in 1846; not seen since 1852
Encke's Comet	1786	3.3	Parent of Taurid meteors; shortest known period of any comet
Comet Swift–Tuttle	1862	130	Parent of Perseid meteors; reappeared 1992
Comet Ikeya–Seki	1965	880	'Sun-grazing' comet, passed 500,000 km/ 300,000 mi above surface of the Sun on 21 October 1965
Comet Kohoutek	1973	–	Observed from space by Skylab astronauts
Comet West	1975	500,000	Nucleus broke into four parts
Comet Bowell	1980	–	Ejected from Solar System after close encounter with Jupiter
Comet IRAS–Araki–Alcock	1983	–	Passed only 4.5 million km/2.8 million mi from the Earth on 11 May 1983
Comet Austin	1989	–	Passed 32 million km/20 million mi from the Earth in 1990
Comet Shoemaker–Levy 9	1993	–	Made up of 21 fragments; crashed into Jupiter in July 1994
Comet Hale–Bopp	1995	1,000	Formed a coma larger than the Sun, largely consisting of carbon monoxide; clearly visible with the naked eye in March 1997
Comet Hyakutake	1996	–	passed 15 million km/9.3 million mi from the Earth in 1996

communications satellite or comsat

An artificial **satellite** in orbit around the Earth, relaying global telephone, television, telex, and other messages. Messages are mostly sent to and from the satellites via ground stations, but direct broadcast satellites are powerful enough to transmit direct to small domestic aerials. The world is now linked by a system of communications satellites. Other satellites are used by individual countries for internal communications, or for business or military use.

- Most communications satellites are in **geostationary orbit**, appearing to hang fixed over one point on the Earth's surface.
- The first satellite to carry TV signals across the Atlantic Ocean was *Telstar* in July 1962.
- The power for a communications satellite is produced by solar cells. A typical comsat needs about 2 kW of power, the same as an electric heater.

communications satellite *The Early Bird satellite.*

conjunction

The alignment of two celestial bodies as seen from the Earth. A conjunction takes place when a **planet** is closely aligned with another celestial object, such as the Moon, a star, or another planet. In particular, one of the bodies may be the Sun.

- A **superior planet** (or other object) is said to be at conjunction when it lies behind the Sun.
- An **inferior planet** (or other object) comes to *inferior conjunction* when it passes between (or almost between) the Earth and the Sun.
- An inferior planet (or other object) is at *superior conjunction* when it passes behind (or almost behind) the Sun.

Because the orbital planes of **Mercury** and **Venus** are tilted with respect to that of the Earth, they usually pass either above or below the Sun at inferior conjunction. If they line up exactly, a *transit* will occur, with the planet briefly appearing as a black dot against the background of the Sun.

constellation

Originally one of the groupings of stars that ancient astronomers devised in order to describe the sky. For today's astronomers, one of 88 areas into which the sky is divided for the purposes of identifying and naming celestial objects. Originally a constellation was regarded as being restricted to the conventional figure, thus leaving many unattached stars unidentified. The first cartographer to draw boundaries between adjacent constellations and fill up the sky within them was the German astronomer Johann Elert Bode (1747–1826) in his *Uranographia* of 1801, with figures based on those of Geman artist Albrecht Dürer (1471–1528). These boundaries became more important than the figures themselves and have now been fixed by international agreement.

Naming of the constellations

c. 2800 BC Possible origin of Western constellations: They were simple, arbitrary patterns of stars in which the Babylonians and other early civilizations visualize gods, sacred beasts, and mythical heroes

AD 150 In the *Almagest*, **Ptolemy** catalogues the positions of 1,022 of the brightest stars in 48 constellations. He leaves a blank area centred on a point that would have been the celestial south pole *c.* 2800 BC.

1515 The German artist Albrecht Dürer (1471–1528) publishes two maps, one for each hemisphere, with the stars plotted by Ptolemy incorporated into appropriate figures. These figures are copied by most subsequent cartographers.

1603 The German astronomer Johann Bayer of Augsburg (1572–1625) publishes his *Uranometria*, using Dürer's figures. He identifies the individual stars

constellation *Part of a chart of the northern skies showing the constellations.*

within each constellation by Greek and Latin letters assigned partly in order of brightness, and partly with regard to their position in the constellation figure. He also gives a chart of the south polar region, formalizing the 12 southern constellations used by early ocean navigators

1751–53 Nicolas de Lacaille (1713–1762) adds 14 fainter constellations to the southern sky as a result of his systematic observations at the Cape of Good Hope in South Africa.

1930 The International Astronomical Union adopts the system still used, in which there are 88 constellation, assigned to areas with straight-line boundaries.

THE CONSTELLATIONS TODAY

In 1930 the International Astronomical Union formally adopted a system in which there are 88 constellations with definitive boundaries. These are lines of constant right **ascension** (north–south) or constant **declination** (east–west), as these were at the time of the spring **equinox** of 1875. In addition to any special name (such as **Polaris**) that a star or other object may have, it is identified by a letter or number followed by the possessive form of the Latin name of its constellation. For example, Polaris is Alpha **Ursae Minoris**, meaning Alpha in the constellation **Ursa Minor**. This is usually contracted to a UMi. ('a' is the Greek letter 'alpha'). The double star Albireo is Beta Cygni, meaning it is star Beta in the constellation **Cygnus**, or b Cyg ('b' is the Greek letter 'beta'.)

Copernican system

A theory of the universe proposed by Nicolaus **Copernicus** in the 16th century, in which the Sun is at the centre of the **Solar System**. This **heliocentric**

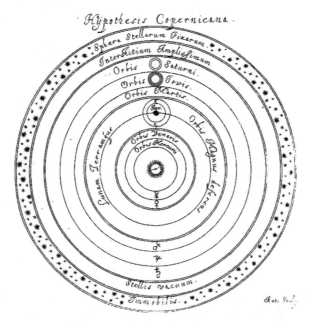

Copernican system *The Copernican (heliocentric) system of the universe.*

(Sun-centred) model eventually replaced the **geocentric** (Earth-centred) model of **Ptolemy**.

Like the Greeks before him, however, Copernicus was able to imagine only circular planetary **orbits**. This forced him to retain Ptolemy's system of epicycles, with the Earth revolving around a centre that revolved around another centre, which in turn orbited the Sun. Copernicus also held to the notion of spheres in which the planets were supposed to travel.

The theory did not immediately have a great effect in the sphere of philosophy and theology. The reaction was favourable among Roman Catholics but antagonistic among Lutherans. There was no question of persecution, and it was only when new philosophies began to develop from it that Copernicus's book was denounced by Luther. It was placed on the Roman Catholic Index of Forbidden Books in 1616, and not removed until 1835.

WHAT COPERNICUS SAID:

- The Earth rotates daily about its axis, thus accounting for the apparent daily rotation of the sphere of the fixed stars.
- A 'wobble' in this rotation of the Earth on its axis accounts for the **precession** of the **equinoxes.**
- The Earth makes one full orbit of the Sun in a year, whereas the Moon orbits the Earth.
- The orbits of the planets **Mercury** and **Venus** are inside the Earth's. This explains why these planets never appear to move far from the Sun.
- **Mars**, **Jupiter**, and **Saturn** travel in orbits outside the Earth's, at a slower pace than the Earth. This explains why these planets sometimes display **retrograde motion** – backward movement – when the Earth overtakes them.

Copernicus, Nicolaus (1473–1543)

Polish astronomer who proposed that the Sun, not the Earth, is at the centre of the **Solar System**. His great work *De Revolutionibus Orbium Coelestium/On the Revolutions of the Heavenly Spheres* was the all-important first step to the more accurate picture of the Solar System built up by later astronomers. Copernicus worked for more than 30 years before agreeing to the publication of this book. This was not, as is often supposed, because he feared that it would be seen as heretical, but because his idea was at that time so incredible that only those with an intimate knowledge of astronomy could be expected to consider it. Pope Paul III accepted the dedication of the work to himself. Andreas Osiander, a Lutheran minister, oversaw the publication and inserted a preface (without permission) stating that the theory was intended merely as an aid to the calculation of planetary positions, not as a statement of reality. This compromised the value of the text in the eyes of many astronomers, but it also saved the book from instant condemnation by the Roman Catholic Church.

COPERNICUS' LIFE

1473	Copernicus (Polish name Mikolaj Kopernik) is born in Torun, on the river Vistula, Poland.
1483	After the death of his father, he is adopted by his uncle, afterwards Bishop of Ermland. He studies mathematics, astronomy, classics, law, philosophy, and medicine at Kraków and various universities in Italy.
1497	He begins to make astronomical observations, although he relies mainly on others' data.
1500	Lectures on mathematics in Rome with great success.
1506	On return to Poland becomes physician to his uncle, the bishop of Varmia, and also canon at Frombork. He is able to intersperse astronomical work with his duties.
***c.* 1513**	Writes a brief, anonymous text entitled *Commentariolus*, outlining his new theory, but it is not published until the 19th century.
1530	Completes his major work, *De Revolutionibus Orbium Coelestium/On the Revolutions of the Heavenly Spheres*, but does not yet publish.
1543	*De Revolutionibus* published; Copernicus dies.

See also: *Copernican system, Galilei, Galileo, Ptolemaic system.*

Coriolis force

A force acting on the surface of a **planet**, resulting from its rotation. On **Jupiter**, for example, it causes the swirling of the storm called the Great Red Spot. On the Earth it affects the atmosphere and all objects on the Earth's surface. It causes winds and ocean currents to be deflected to the right in the northern hemisphere; in the southern hemisphere it causes deflection to the left. The effect is named after its discoverer, the French mathematician Gaspard de Coriolis (1792–1843).

corona

A faint halo of hot gas around the Sun. It is very hot – about 2,000,000°C/3,600,000°F – but very tenuous (thin). It is visible during solar **eclipses** or through a **coronagraph**. Gas flows away from the corona to form the **solar wind**.

Corona Australis or Southern Crown
A small **constellation** of the southern hemisphere, located near the constellation **Sagittarius**. It is similar in size and shape to **Corona Borealis** but is not as bright.

Corona Borealis or Northern Crown
A small but easily recognizable **constellation** of the northern hemisphere, between **Hercules** and **Boötes**. Its brightest star is Alphecca (or Gemma), which is 78 **light years** from Earth.

Corona Borealis contains several variable stars:

- R Coronae Borealis is normally fairly constant in brightness but fades at irregular intervals and stays faint for a variable length of time.

- T Coronae Borealis is normally faint, but very occasionally blazes up and for a few days may be visible to the naked eye. It is a recurrent **nova**.

The constellation is traditionally identified with the jewelled crown of the Cretan princess Ariadne, which in Greek mythology was cast into the sky by the god Dionysus.

coronagraph
A device for studying the solar **corona** at any time of the day. It was first invented by the French astronomer Bernard Lyot (1897–1952).

cosmic background radiation
Electromagnetic radiation filling space and left over from the original formation of the universe in the **Big Bang** around 15 billion years ago. It corresponds to an overall background temperature of 3K (–270°C/–454°F), or 3°C above absolute zero. In 1992 the Cosmic Background Explorer satellite, **COBE**, detected slight 'ripples' in the strength of the background radiation that are believed to mark the first stage in the formation of **galaxies**. The cosmic background radiation was first detected in 1964 by US physicists Arno Penzias (1933–) and Robert Wilson (1936–), who in 1978 shared the Nobel Prize for Physics for their discovery.

cosmic radiation
Streams of high-energy particles arriving at the Earth's surface from space, consisting of **protons**, **alpha particles**, and light nuclei. Some reach the Earth directly, others collide with atomic nuclei in the Earth's **atmosphere**, and produce secondary nuclear particles (chiefly mesons, such as pions and muons) that shower the Earth and are also called cosmic radiation. Those of low energy seem to originate in our **Galaxy**, while those of high energy seem to be of extragalactic origin. The galactic particles may come from

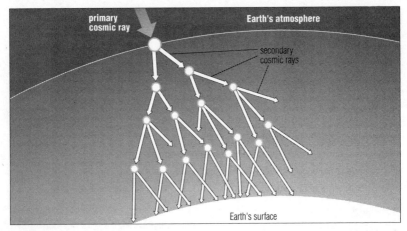

cosmic rays *Primary cosmic rays consisting of protons, alpha particles, and light nuclei collide with nuclei in the Earth's atmosphere to produce secondary cosmic rays called cosmic radiation.*

supernova explosions or **pulsars**. At higher energies, other sources are necessary, possibly the giant jets of gas which are emitted from some **galaxies**.

cosmic string
A hypothetical 'rope' of energy trapped since the **Big Bang**, crossing the entire universe and with a colossal mass of trillions of tonnes per centimetre. Certain theories of the birth of the universe predict the existence of cosmic strings. If they exist, they will show themselves by their **gravitational lensing** effect on the light of **galaxies** lying beyond them. However, no trace of them has yet been found.

cosmological constant
A number describing the strength of a hypothetical repelling force driving apart all matter in the universe. Such a force was briefly proposed by the physicist Albert **Einstein** because, when combined with his general theory of **relativity**, it predicted a stable universe – one that was neither expanding nor contracting. Einstein assumed that the universe must be this way. When the expansion of the universe was discovered by Edwin **Hubble** in 1929, Einstein abandoned the idea of such a force, calling it 'the greatest mistake of his life'. But recent studies of **supernovae** in distant **galaxies** suggest that there may be such a force (that is, the cosmological constant may be non-zero). This means that the galaxies may be slowing down less rapidly than they otherwise would, in fact, they may even be accelerating.

See also: *Big Bang, red-shift.*

Crab nebula

A cloud of gas 6,000 **light years** from Earth, in the constellation **Taurus**. It is the remnant of a **supernova**, or exploding star. The **nebula** is a powerful radio and **X-ray** source. Optically it appears as a diffuse elliptical area on which is superimposed an intricate network of bright filaments. Its diameter is about 10 light years. Its light is highly polarized, suggesting the presence of strong **magnetic fields**. The diffuse portion emits radiation throughout the whole electromagnetic **spectrum** from radio to **gamma rays**.

CRAB NEBULA FACTS

- The nebula was named by English-born Irish astronomer William Parsons Rosse (1800–1867) from its crablike shape.
- It is the remnant of a star that, according to Chinese astronomical records, was observed as a brilliant point of light on 4 July 1054. This 'new star' was actually an old, massive star exploding as a supernova at the end of its life.
- The Crab Nebula is still increasing in size from the force of the explosion.
- The energy given out by the nebula comes from a **pulsar** near its centre that flashes 30 times a second.

crater

A bowl-shaped depression in the surface of the Earth or some other celestial body, usually round and with steep sides. **Craters** are formed by explosive events such as the eruption of a **volcano**, the explosion of a bomb, or the impact of a **meteoroid**. The Moon has more than 300,000 craters over 1 km/0.6 mi in diameter, formed by meteoroid bombardment; similar craters on Earth have mostly been worn away by erosion. Craters are found on many other bodies in the **Solar System**, including **Mercury**, **Mars**, and all natural **satellites** and **asteroids**. Craters

The US lunar probe Clementine discovered an enormous crater on the far side of the Moon in 1994. The South Pole-Aitken crater is 2,500 km/1,563 mi across and 13 km/8 mi deep, making it the largest known crater in the Solar System.

produced by impact have different shapes from those formed by volcanic activity, enabling astronomers to distinguish likely methods of crater formation on planets in the Solar System. Unlike volcanic craters, impact craters have a raised rim and central peak and are almost always circular, irrespective of the meteoroid's angle of fall.

See also: *caldera*.

Crux

A **constellation** of the southern hemisphere, popularly known as the Southern Cross. It is the smallest of the 88 constellations but one of the brightest. Its brightest stars are Alpha Crucis (or **Acrux**), a double star about 400 **light years** from Earth, and Beta Crucis (or Mimosa).

- Crux is one of the best known constellations, as it is represented on the flags of Australia and New Zealand.

- Near Beta Crucis lies a glittering **open cluster** known as the Jewel Box, so named by John **Herschel**.

- The constellation also contains the Coalsack, a dark **nebula** silhouetted against the bright starry background of the **Milky Way**.

Cygnus

A large and prominent **constellation** of the northern hemisphere, represented as a swan. Its brightest star is Alpha Cygni, or **Deneb**, which is of the first **magnitude**. Beta Cygni (Albireo) can be seen through small **telescopes** to be a yellow and blue double star. The area is rich in high-luminosity objects, **nebulae**, and clouds of obscuring matter. Deneb marks the tail of the swan, which is depicted as flying along the **Milky Way**. Some of the brighter stars form the Northern Cross, the upright being defined by Alpha, Gamma, Eta, and Beta, and the crosspiece by Delta, Gamma, and Epsilon Cygni.

CYGNUS CONTAINS MANY INTERESTING OBJECTS:

- The North America nebula is named from its shape.

- The Veil nebula is part of an enormous ring of nebulosity, the remains of a **supernova** that exploded about 50,000 years ago.

- Cygnus A is a powerful radio source, and was the first celestial radio source other than the Sun to be discovered. It is apparently a double **galaxy**.

- The **X-ray** source Cygnus X-1 is thought to mark the position of a **black hole**.

- One of the fainter stars, 61 Cygni, has an exceptionally large **proper motion** – that is, it seems to move fast across our sky. The German astronomer Friedrich Bessel (1784–1846) realized this meant it is probably close to us, and in 1838 used it for his first measurement of a stellar distance. He found it to be only 10.4 light years from the Sun.

dark matter

Matter that might make up 90–99% of the mass of the universe – yet so far remains undetected. Some current theories of cosmology suggest that there is at least 10 times as much matter in the universe as can be observed. The existence of this **missing mass** would explain many currently puzzling gravitational effects in the movement of **galaxies**. Some dark matter could exist between the galaxies in the form of dark gas, or within galaxies as **brown dwarfs**, which are stars that have failed to light up. In 1993 astronomers identified part of the dark matter. These objects are known as MACHOs (massive astrophysical compact halo objects) and may make up approximately half of the dark matter in the **Milky Way's** halo. If **neutrinos**, which fill all space, have a mass, they might account for some of the dark matter. The universe might also be filled with a sea of particles called WIMPs (weakly interacting massive particles), a completely new kind of matter.

dark nebula

A **nebula**, or gas cloud, that does not give out light of its own, or shine by reflected light. Dark nebulae consist largely of ordinary **hydrogen**, sometimes mixed with **dust** particles. They reveal their presence when they blot out the light of stars behind.

day

The time taken for the Earth to rotate once on its axis. It has slightly different values according to the definition of a complete rotation.

- The **sidereal** day is the time that the Earth takes to rotate once relative to the stars. It is has a practically constant length of 23 hours 56 and 4.1 seconds.

- The *solar day* is the time that the Earth takes to rotate once relative to the Sun. It is longer than the sidereal day, because the Sun's position against the background of stars as seen from Earth changes as the Earth moves. The solar day also varies in length as the Earth moves around its **orbit**. Its average length is defined as being exactly equal to 24 hours, and is the basis of our civil day.

declination
The north–south angle between a celestial object and the celestial equator. It is the coordinate on the **celestial sphere** (imaginary sphere surrounding the Earth) that corresponds to **latitude** on the Earth's surface. Declination runs from 0° at the celestial equator to 90° at the north and south celestial poles.

Deimos
The smaller of the two moons of **Mars**. It is irregularly shaped, but is not as heavily cratered as the other moon, **Phobos**. Deimos was discovered in 1877 by a US astronomer, Asaph Hall (1829–1907), and is thought to be an **asteroid** captured by the gravity of Mars.

DEIMOS: STATISTICS

Diameter	Distance from centre of planet	Period
15 x 12 x 11 km / 9 x 7.5 x 7 mi	23,500 km/14,600 mi	1.26 days

Deneb or Alpha Cygni
The brightest star in the constellation **Cygnus**, and the 20th-brightest star in the night sky. It is one of the greatest **supergiant** stars known, with a true luminosity about 60,000 times that of the Sun. Deneb is about 1,800 **light years** from the Sun.

deuterium
Naturally occurring **isotope** of **hydrogen** – that is, a form of hydrogen that is chemically identical to ordinary hydrogen but whose atom has a different mass. It is sometimes given the symbol D.

• The deuterium atom contains one **proton** and one **neutron** in its nucleus and was discovered in 1932.

• In nature, about one in every 6,500 hydrogen atoms is deuterium.

• Most of the deuterium in the universe was formed in the first few seconds following the **Big Bang**, and it features in the **proton–proton cycle** that powers the Sun and similar stars.

diamond ring
A beautiful optical effect in a **total eclipse** produced by the last thin arc of the Sun to be seen before it is completely hidden.
   See also: *Bailey's beads.*

diffraction

The spreading out of waves when they pass through a small gap or around a small object. In order for this effect to be observed the size of the object or gap must be comparable to or smaller than the wavelength of the waves. Diffraction occurs with all types of wave – electromagnetic radiations, sound waves, and waves in water. It explains such phenomena as why sound is heard round corners and why radio waves of long wavelength can bend round hills better than those of short wavelength.

- Diffraction of **light** causes the formation of rings and rays around **telescope** images of stars. It means that the diameters of stars cannot be measured directly.

- A diffraction grating is a plate of glass or metal ruled with closely spaced parallel lines. It separates light waves or other radiations into their component wavelengths to form a **spectrum**.

   See also: *interferometry*.

direct motion or prograde motion

1 Apparent movement of a planet from west to east in relation to the background of stars. Occasionally the **superior planets** appear to reverse their motion and move east to west in **retrograde motion**.

2 Movement of a **satellite** around a planet counterclockwise as viewed from above the planet's north pole. Most satellites move in this sense, but a few are retrograde.

dispersion

The separation of waves of different frequency in a particular medium because the speed of the waves depends on their frequency (or wavelength). In the case of visible **light** the frequency corresponds to colour.

- The splitting of white light into a **spectrum** when it passes through a **prism** occurs because each component frequency of light moves through at a slightly different angle and speed.

- A rainbow is formed when sunlight is dispersed by raindrops.

- Astronomers learn about interstellar gas by, among other things, its dispersive effects on light and radio waves.

Doppler effect

The change in the observed frequency (or wavelength) of waves due to relative motion between the source of waves and the observer. The Doppler effect is responsible for the perceived change in pitch of a siren as it approaches and then recedes, and for the **red-shift** of light from distant galaxies (which are all receding from us). It is named after the Austrian physicist Christian Doppler (1803–1853).

Draco
A large but faint **constellation** represented as a dragon coiled around the north celestial pole.

- The star Alpha Draconis (Thuban) was the pole star 4,800 years ago. The position of the celestial pole has changed because of **precession**.

- Thuban seems to have faded, for it is no longer the brightest star in the constellation as it was at the beginning of the 17th century. Gamma Draconis is more than a **magnitude** brighter. It was extensively observed by James Bradley (1693–1762), who from its apparent changes in position discovered the **aberration** of starlight and **nutation**.

dust
Dust is a major component of the matter in interplanetary and interstellar space. **Meteoroids** and **micrometeoroids** are dust grains. The tails of **comets** largely consist of streams of dust, and **planetary rings** largely or wholly consist of dust. The dark 'lanes' seen in the **Milky Way** are dust clouds blocking out the starry background beyond. Even where interstellar dust clouds are not evident, they dim the light from distant stars, preventing us from seeing all the way to the centre of our Galaxy. The dust affects the plane of polarization of starlight, giving evidence of the shape and size of the dust grains. Whereas **elliptical galaxies** are relatively free of dust, the arms of **spiral galaxies** are rich in dust from which new stars are constantly being born.

See also: *Gegenschein.*

dwarf galaxies
Small **elliptical galaxies**, having masses of only a million times that of the Sun or less. They are the most common type of **galaxy**.

Earth

The third **planet** from the Sun. It is almost spherical, though flattened slightly at the poles, and is composed of three concentric layers: the core, the mantle, and the crust. About 70% of the surface (including the north and south polar ice caps) is covered with water. The Earth is surrounded by a life-supporting **atmosphere**, uniquely rich in oxygen, and is the only planet on which life is known to exist.

Earth's size, shape and motion

Mean distance from the Sun	149,600,000 km/92,960,000 mi
Equatorial diameter	12,756 km/7,926 mi
Polar diameter	12,714 km/7,900 mi
Equatorial circumference	40,074 km/24,901 mi
Polar circumference	39,995 km/24,852 mi
Rotation period	23 hr 56 min 4.1 sec
Year (complete orbit, or sidereal period)	365 days 5 hr 48 min 46 sec
Average speed around the Sun	30 kps/18.5 mps
Tilt of equator relative to plane of orbit	23.5°

Earth's atmosphere

Nitrogen	78.09%
Oxygen	20.95%
Argon	0.93%
Carbon dioxide	0.03%
Neon, helium, krypton, hydrogen, xenon, ozone, radon	Less than 0.0001%

Earth's surface

Land area	150,000,000 sq km/57,500,000 sq mi
Greatest height above sea level (Mount Everest)	8,872 m/29,118 ft
Ocean area	361,000,000 sq km/139,400,000 sq mi
Greatest depth (Mariana Trench, Pacific)	11,034 m/36,201 ft

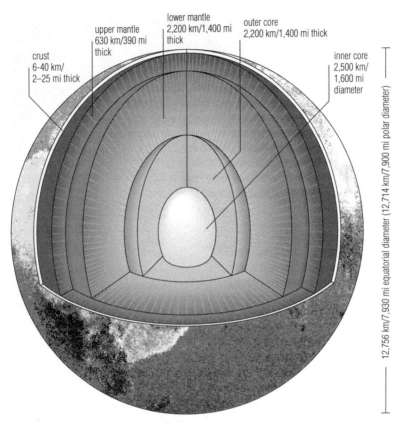

Earth *The surface of the Earth is a thin crust about 6 km/4 mi thick under the sea and 40 km/25 mi thick under the continents. Under the crust lies the mantle about 2,900 km/1,800 mi thick and with a temperature of 1,500–3,000°C/2,700–5,000°F. The outer core is about 2,200 km/1,400 mi thick, of molten iron and nickel. The inner core is probably solid iron and nickel at about 5,000°C/9,000°F.*

Drifting continents

The crust and the topmost layer of the mantle, called the lithosphere, form about twelve major sections, or plates. Some of these carry the continents. The plates are in constant, slow motion, called tectonic drift. They are able to move because the underlying layer of the mantle, the asthenosphere, is weaker and more elastic than the surrounding mantle. The top of the asthenosphere lies at a depth of approximately 100 km/60 mi and the bottom at approximately 260 km/160 mi. The elastic and relatively fluid asthenosphere allows the overlying, more rigid plates of the lithosphere to move sideways in a process known as plate tectonics. It also allows the overlying crust and mantle to move vertically in response to **gravity**, so the continents float in the mantle as ships float in water.

A difference of spin
US geophysicists announced in 1996 that they had detected a difference in the spinning time of the Earth's core and the rest of the planet; the core is spinning slightly faster.

Earth viewed from the Moon.

The birth of the Earth
The Earth was formed with the rest of the **Solar System** 4.6 billion years ago by consolidation of **dust** in the cloud of matter that circled the newly formed Sun. Life appeared on Earth 3.5–4 billion years ago

---

To remember the most common elements of the planet's crust, in descending order of abundance:

only silly asses in college study past midnight are the initial letters of the names of the elements:
oxygen, silicon, aluminium, iron, calcium, sodium, potassium, magnesium

---

Earth resources satellite
An artificial **satellite** that surveys the surface of the Earth and gathers information about natural resources, such as bodies of water, types of vegetation, geological formations, and so on. Such satellites have relatively low orbits, and identify types of surface by their differing reflectivity at different wavelengths.

See also: *spy satellite.*

eclipse

The darkening of a celestial object because it passes into shadow (as when the Moon passes into the Earth's shadow) or because its own light is blocked (as when the Sun's light is blocked by the Moon). The term is usually used for **solar** and **lunar eclipses**, which may be either partial or total, but may also refer to other bodies – for example, to an eclipse of one of Jupiter's moons by **Jupiter** itself. An eclipse of a star by a body in the **Solar System** is also called an **occultation**.

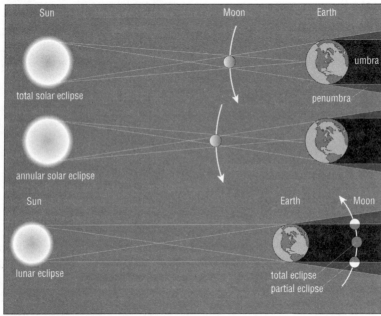

eclipse *The two types of eclipse: lunar and solar. A lunar eclipse occurs when the Moon passes through the shadow of the Earth. A solar eclipse occurs when the Moon passes between the Sun and the Earth, blocking out the Sun's light. During a total solar eclipse, when the Moon completely covers the Sun, the Moon's shadow sweeps across the Earth's surface from west to east at a speed of 3,200kph/2,000mph.*

Solar and lunar eclipses in 2000–03

Month	Day	Type of eclipse	Duration of maximum eclipse	Region for observation
2000				
January	21	lunar total	4 hr 44 min	the Americas, Europe, Africa, western Asia

Solar and lunar eclipses in 2000–03 (continued)

Month	Day	Type of eclipse	Duration of maximum eclipse	Region for observation
February	5	solar partial	12 hr 50 min	Antarctica
July	1	solar partial	19 hr 34 min	southeastern Pacific Ocean
July	16	lunar total	13 hr 56 min	southeastern Asia, Australasia
July	31	solar partial	2 hr 14 min	Arctic regions
December	25	solar partial	17 hr 36 min	USA, eastern Canada, Central America, Caribbean
2001				
January	9	lunar total	20 hr 21 min	Africa, Europe, Asia
June	21	solar total	12 hr 4 min	central and southern Africa
December	14	solar annular	20 hr 52 min	Pacific Ocean
2002				
June	10	solar annular	23 hr 44 min	Pacific Ocean
December	4	solar total	7 hr 31 min	southern Africa, Indian Ocean, Australia
2003				
May	16	lunar total	3 hr 40 min	the Americas, eastern and western Africa
May	31	solar annular	4 hr 8 min	Iceland, Greenland
November	9	lunar total	1 hr 18 min	the Americas, Africa, Europe
November	23	solar total	22 hr 49 min	Antarctica

eclipsing binary

A **binary** (double) star in which each star periodically **eclipses** the other (passes in front of it) as seen from the Earth. When one star crosses in front of the other, the total light received on Earth from them decreases. The first eclipsing binary to be noticed was **Algol**.

ecliptic

The path that the Sun appears to follow throughout the year against the background of the stars. It is really due to the motion of the Earth around the Sun. It can be thought of as the plane of the Earth's **orbit** projected on to the **celestial sphere** (the imaginary sphere around the Earth). The ecliptic is tilted at about 23.5° with respect to the celestial equator, a result of the tilt of the Earth's axis relative to the plane of its orbit around the Sun.

See also: *seasons, zodiac.*

ecliptic coordinates

A system for measuring the position of astronomical objects on the **celestial sphere**. Position is measured in relation to the **ecliptic**, the apparent path of the Sun around the sky during the year.

- Ecliptic latitude (symbol β, which is the Greek letter beta) is measured in degrees from the ecliptic (= 0°) to the north (= 90°) and south (=−90°) ecliptic poles.

- Ecliptic longitude (symbol λ, which is the Greek letter lambda) is measured in degrees eastward along the ecliptic (= 0° to 360°) from a fixed point known as the **first point of Aries** or the vernal **equinox**.

- Ecliptic **latitude** and **longitude** are sometimes known as celestial latitude and longitude. The ecliptic longitude of the Sun (solar longitude) is a convenient measure of the position of the Earth in its orbit.

Effelsberg radio telescope
The world's largest fully steerable **radio telescope**, located near Bonn, Germany. It is a dish 100 m/328 ft across. It began work in 1971 and is operated by the Max Planck Institute for Radio Astronomy.

EGG
Acronym for evaporating gaseous globule, a comparatively dense knot of gas and **dust** in a region of newly forming stars. Stars are born when **nebulae** (giant clouds of dust and gas) contract under the influence of **gravity**. These clouds consist mainly of **hydrogen** and **helium**, with traces of other elements and dust grains. A huge volume of **interstellar matter** gradually separates from the cloud, and the temperature and pressure in its core rises as the star grows smaller and denser. As the star is forming, it is surrounded by EGGs, which were first photographed in the Eta Carina Nebula in 1996 by the **Hubble Space Telescope**.

Einstein, Albert (1879–1955)
German-born physicist whose two theories of **relativity** revolutionized our understanding of matter, space, and time. All astronomical and cosmological work that involves strong gravitational fields, or thermonuclear processes, depends on Einstein's ideas. All modern theorizing about the origin and destiny of the universe is carried out on the basis of his general theory of relativity. He also transformed other areas of physics. He established that **light** has a particle nature, in work for which he was awarded the Nobel Prize for Physics in 1921. He also investigated Brownian motion, confirming the existence of atoms.

Work relevant to astronomy and cosmology
By 1905 there had been a steady accumulation of experimental results suggesting that light and other form of electromagnetic radiation do not behave as predicted by classical physics. It proved impossible, for example, to

measure the expected changes in the speed of light relative to the motion of the Earth. Einstein recognized that the speed of light is independent of the speed of the observer. It followed that time and distance must vary, depending on the velocity of each observer.

Einstein deduced that, in a system in motion relative to an observer, length would be observed to decrease, time would slow down, and mass would increase. The magnitude of these effects is negligible at ordinary velocities. But as a spaceship approached the velocity of light, its length relative to an observer at rest would shrink, time as indicated by onboard clocks would slow, and the mass of the system would increase without limit. Einstein concluded that no system

Einstein *Albert Einstein, German-born US physicist whose work on relativity underpins modern astronomy and cosmology involving strong gravitational fields.*

can actually reach or exceed the velocity of light. Einstein's ideas were later verified with observations of fast-moving subatomic particles and atomic clocks transported around the world. In 1907 Einstein went on to show that mass is related to energy by the famous equation: $E = mc^2$.

In his general theory of relativity, Einstein viewed the properties of **space-time** as being altered by the presence of a body with mass. A **planet's** orbit around the Sun arises from the Sun's alteration of space-time in its region. Furthermore, light-rays should bend when

The equation $E=mc^2$ states that a mass m can be converted into an amount of energy E equal to the mass multiplied by the speed of light (c) squared, a truly enormous number.

they pass by a massive object such as the Sun. As a result, the apparent positions of stars would shift when they are seen near the Sun. Einstein was triumphantly vindicated when observations of a **solar eclipse** in 1919 showed apparent shifts of approximately the amount he had predicted.

For the rest of his career Einstein strove to combine the theories of electromagnetism, gravitation and two newly discovered nuclear forces into a 'unified field theory', but with little success. Such a unification is essential to understanding what happened in the first instance after the birth of the universe in the **Big Bang**, before the forces had separated from each other. Later physicists have made limited progress, and a major aim of modern physics remains to unify all four fundamental forces, thereby achieving Einstein's dream.

See also: *cosmological constant.*

electron

A subatomic particle that is a constituent of all atoms. Electrons belong to the class of particles known as leptons, together with **neutrinos** and tau-particles.

- The electron carries a charge of 1.602192×10^{-19} coulomb.

- It has a mass of 9.109×10^{-31} kg, which is $1/1,836$ times the mass of a **proton**.

- The electrons in each atom surround the nucleus in groupings called shells; in a neutral atom the number of electrons is equal to the number of protons in the nucleus.

- The electron shell structure is responsible for the chemical properties of the atom.

- Electric currents are streams of electrons that have become separated from atoms.

- A beam of electrons undergoes **diffraction** (scattering) and produces interference patterns in the same way as electromagnetic waves such as **light**; hence they may be regarded as waves as well as particles.

elements, abundance of

By far the most common element in the universe is **hydrogen**, which has the simplest atom, consisting of a single **proton** circled by a single **electron**. **Helium**, with two protons and two **neutrons** in its nucleus, comes next. Generally speaking the abundance of elements in our

Abundance of the elements

For every 1 million hydrogen atoms there are the following numbers of atoms of other elements. Only the most common elements are listed

Element	Symbol	No. of atoms
Helium	He	50,000
Oxygen	O	670
Carbon	C	350
Nitrogen	N	110
Silicon	Si	35
Magnesium	Mg	34
Neon	Ne	28
Sulphur	S	6
Calcium	Ca	2

universe decreases the heavier and more complex their atoms are. The hydrogen, helium, lithium and some other light atoms were built up from subatomic particles in the **Big Bang**. Other atoms were built up from lighter atoms in the centres of stars. Some of these elements were scattered through the universe when the most massive stars exploded as **supernovae**.

The elements found on Earth are not representative of the universe as a whole: oxygen, combined with other elements in the rocks, is the most common element here, while hydrogen is rare. Scientists learn about the composition of the rest of the universe by analysing starlight by means of **spectroscopy** and analysing the composition of **meteorites**.

ellipse
A closed curve resembling a foreshortened ('squashed') circle. The adjective is 'elliptical'.

Planets move around the Sun, and **satellites** around planets, in **orbits** that are elliptical.

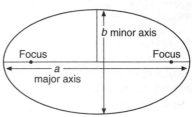

ellipse *The main features of an ellipse.*

- An ellipse is one of a series of curves known as conic sections, because it can be produced by making a slice across a cone that does not pass through the base. (If the slice is parallel to the base it will produce a circle, which is a special kind of ellipse.)

- The longest diameter of an ellipse is called the major axis. The diameter bisecting this at right angles is the minor axis.

- There are two special locations within an ellipse, called its *foci*. For any point, the sum of the distances from the foci is a constant.

- The Sun lies at one of the foci of each planet's orbit.

elliptical galaxy
One of the main types of **galaxy**. Such a galaxy has an outline that is an **ellipse**; its solid shape is an ellipsoid, resembling a rugby football or an American football.

- Unlike **spiral galaxies**, elliptical galaxies have very little gas or **dust** and practically no stars are being formed within them.

- Elliptical galaxies range in size from giant ellipticals, which are often found at the centres of clusters of galaxies and may be strong radio sources, to tiny **dwarf ellipticals**, containing about a million stars.

- More than 60% of known galaxies are elliptical.

elongation
The angle between the Sun and a **planet** or other **Solar System** object, as seen from the Earth. This angle is 0° at **conjunction** and 180° at **opposition**.

Encke division
A narrow gap in the rings of **Saturn** reported in 1838 by the German astronomer Johann Encke (1791–1865). Only 300 km/186 mi wide, its existence was not generally accepted for another 50 years, until it was confirmed by the US astronomer James Keeler (1857–1900).
    See also: *planetary rings.*

ephemeris
A table showing the movements and appearances of celestial bodies over some period of time (plural ephemerides). The governments of many countries print ephemerides for the use of navigators.

equation of time
A mathematical formula which shows the difference between apparent **solar time** and mean solar time throughout the year. The Sun does not move around the **ecliptic** at a constant rate throughout the year, because of variations of the Earth's speed in its orbit. So astronomers calculate mean solar time, which relates to the position of an imaginary *mean Sun* which moves at a constant rate. The equation of time shows the difference between them. The difference is greatest in early November, when the Sun is more than 16 minutes 'fast' compared with the mean Sun.
    See also: *Greenwich mean time, time measurement.*

equinox
Either of the two points at which the **ecliptic**, or annual path of the Sun, crosses the celestial equator; and also either of the times at which the Sun passes through these points.

- The vernal, or spring, equinox occurs about 21 March, with the Sun in the constellation **Pisces**, moving northward.

- The autumnal equinox occurs about 23 September, with the Sun in the constellation **Virgo**, moving southward.

- At the equinoxes, day and night are of equal length.

    See also: *celestial sphere, solstice.*

Eros
An **asteroid**, discovered in 1898, that can pass very close to the Earth. In 1975 it came as close as 22 million km/14 million mi to the Earth. Eros was

the first asteroid to be discovered that has an **orbit** coming within that of **Mars**. It is elongated, and rotates around its shortest axis every 5.3 hours. The **NEAR** (Near-Earth Asteroid Rendezvous) mission was launched in February 1996 on a three-year journey to Eros. It was planned that it would spend a year circling the asteroid in an attempt to determine what it is made of. The initial attempt at a rendezvous in December 1998 failed and the encounter was rescheduled to begin in February 2000.

EROS: STATISTICS

Diameter	Rotation period	Distance from Sun
36 x 12 km / 22 x 7 mi	5.3 hours	218,000,000 km /135,500,000 mi

---

eruptive variable
A star that changes in size and brightness at irregular intervals. Eruptive variables include long-period variables, such as the **red giant Mira** in the constellation **Cetus** (period about 330 days), and irregular variables, such as some red **supergiants**. Eruptive variables emit sudden outbursts of light. Some suffer surface explosions called **flares**, while others, such as **novae**, result from transfer of gas between a close pair of stars.

escape velocity
The minimum velocity that must be given to an object to throw it from the surface of some **planet**, **satellite** or other celestial body in order for it to escape from the pull of its **gravity** without ever needing any further impulse. In the case of the Earth, the escape velocity is 11.2 kps/6.9 mps. Once an interplanetary craft has been boosted to this speed, it will travel on through space indefinitely. (However, its engines will need to be fired again for manoeuvring in space or landing on some body.)

Comparative escape velocities

	Mercury	Venus	Earth	Moon	Mars
Escape velocity	4.2 kps/ 2.6 mps	10.3 kps/ 6.4 mps	11.2 kps/ 7.0 mps	2.38 kps/ 1.5 mps	5.0 kps/3.1 mps

	Jupiter	Saturn	Uranus	Neptune	Pluto
Escape velocity	61.0 kps/ 37.9 mps	37.0 kps/ 23.0 mps	22.0 kps/ 13.7 mps	25.0 kps/ 15.5 mps	1.1 kps/ 0.7 mps

Europa

The fourth-largest moon of the planet **Jupiter**. It is covered by ice and criss-crossed by thousands of thin lines. **NASA's** unmanned *Galileo* **mission** began circling Europa in February 1997. One of the first discoveries was that what had appeared to be cracks covering the surface of Europa are in fact low ridges. Further investigation is needed to determine their origin. NASA announced plans to launch *Europa Observer* in 2003. Its aim will be to search for water beneath Europa's icy surface, where it is possible that some simple form of life may have developed.

EUROPA: STATISTICS

Diameter	Distance from centre of planet	Period
3,140 km/1,950 mi	671,000 km/417,000 mi	3.55 days

European Southern Observatory

Major astronomical **observatory** with two sites on the mountain peaks of Chile, at La Silla and Paranal. It is operated jointly by a consortium of countries that includes Belgium, Denmark, France, Germany, Italy, the Netherlands, Sweden, and Switzerland, with headquarters at Garching, Germany. The principal instrument at La Silla is the **New Technology Telescope**; the main one at Paranal is the **VLT** (Very Large Telescope), consisting of four units that will work together to give unprecedented detail by means of **interferometry**.

European Space Agency (ESA)

Organization of European countries (including Austria, Belgium, Denmark, Finland, France, Germany, Ireland, Italy, the Netherlands, Norway, Spain, Sweden, Switzerland, and the UK) that engages in space research and technology. It was founded in 1975, with headquarters in Paris. ESA has developed various scientific and communications **satellites**, the *Giotto* **mission**, and the *Ariane* rockets. ESA built **Spacelab**, and plans to build its own space station, *Columbus*, for attachment to the **International Space Station**. The ESA's Earth-sensing satellite *ERS*-2 was launched successfully in 1995. It will work in tandem with *ERS*-1, which was launched in 1991, and should improve measurements of the **ozone layer**.

event horizon

The 'edge' of a **black hole**; the surface surrounding the black hole from within which no matter or radiation can escape.

exobiology

The study of life forms that may possibly exist elsewhere in the universe. Exobiologists have a wide range of techniques available to them. They look for signs of past life in **meteorites** that are believed to have reached the Earth from the Moon or **Mars**. Space probes search for organic molecules on other **planets** and moons in the **Solar System**, and in **nebulae**. In **SETI**, the search for extraterrestrial intelligence, **radio telescopes** scan the skies for signals from planets circling other stars.

See also: *origin of life.*

exosphere

The region lying above the Earth's **thermosphere**, where the last traces of atmospheric gas tail off into interplanetary space.

Explorer missions

A series of US scientific **satellites**. *Explorer 1*, launched in January 1958, was the first American satellite in **orbit** and discovered the Van Allen **radiation belts** around the Earth.

faculae

Bright areas on the face of the Sun, commonly in the vicinity of **sunspots**. They were named by the Polish astronomer Johannes Hevelius (1611–1686). They are thought to be caused by luminous **hydrogen** clouds close to the **photosphere**. They last on average about 15 days.

First Point of Aries

Another name for the vernal **equinox**, one of the two points where the **ecliptic** intersects the celestial equator.

    See also: *celestial sphere.*

flare star

A dim red dwarf star that suddenly lights up with great – but brief – luminosity, corresponding to an equally powerful but short-lived burst of **radio** emission. The cause is thought to be a sudden and intense outburst of radiation on or above the star's surface.

flare, solar

A brilliant eruption on the Sun above a **sunspot**, thought to be caused by release of magnetic energy. Flares reach maximum brightness within a few minutes, then fade away over about an hour.

DISRUPTIVE PARTICLES

Flares eject bursts of atomic particles into space at up to 1,000 kps/600 mps.

When these particles reach Earth they enter the radiation belts, and cause them to 'overflow', dumping particles into the atmosphere, with radio blackouts, disruptions of the Earth's magnetic field, and aurorae.

---

flash spectrum

A brief reversal of the **spectrum** of the Sun seen when a **solar eclipse** reaches totality: the bright background of the spectrum becomes dark, while the dark **Fraunhofer lines** become bright. These lines represent wavelengths at which light is absorbed by the cool gases of the Sun's

atmosphere. However, these lines are not completely dark. When the disc of the Sun is covered by the Moon, the bright spectrum disappears, while the Sun's atmosphere can still be seen shining around the Moon's disc. The wavelengths corresponding to the Fraunhofer lines now show up as comparatively bright.

Fomalhaut or Alpha Piscis Austrini

The brightest star in the southern **constellation** Piscis Austrinus and the 18th-brightest star in the night sky. It is 22 **light years** from the Sun, with a true luminosity 13 times that of the Sun.

- Fomalhaut was one of the four 'royal stars' of ancient Persia, marking the approximate positions of the Sun at the **equinoxes** and **solstices**; the other three were **Aldebaran**, **Regulus**, and **Antares**.

- Fomalhaut is one of a number of stars around which *IRAS* (the Infra-Red Astronomy Satellite) detected excess **infrared radiation**, presumed to come from a disc of solid particles around the star. This material may be a **planetary system** in the process of formation.

Foucault pendulum

An enormously long pendulum used to demonstrate the rotation of the Earth. It consists of a long wire to which is attached a heavy weight, which is then free to swing in any direction. Once it has been set swinging in one particular direction, its direction of swing is seen to rotate slowly, turning through 360° in 24 hours. This is in reality the result of the rotation of the Earth beneath the pendulum. This demonstration was first made by the French physicist Léon Foucault (1819–1868) in 1851.

Fraunhofer lines

Dark lines crossing the solar **spectrum**. They are caused by the absorption of **light** from the Sun's bright surface, or **photosphere**, by gases in the cooler, outer regions. They are the key to discovering the chemical elements that are present in the Sun. They were first investigated by the German physicist Joseph von Fraunhofer (1787–1826).

See also: *flash spectrum, spectroscopy.*

free fall

Falling freely under the influence of **gravity**, as a skydiver does in freefall parachuting, or as a spacecraft does when its engines are not firing. In a vacuum, a freely falling body near the Earth's surface accelerates at a rate of approximately 9.807 metres per second per second/32.174 feet per second per second; the value varies slightly at different latitudes, being greatest at the poles and least at the equator. A body falling through air experiences resistance; it accelerates until it reaches a maximum speed called the terminal velocity, when the drag of the air just equals its weight; thereafter, there is no further acceleration.

In **orbit**, astronauts and spacecraft are still held by gravity and are in fact falling freely towards the Earth. Because the craft and their occupants fall at the same rate, they seem weightless in relation to each other. The amount by which an object in a circular orbit falls towards the Earth just equals the amount by which the Earth's surface curves away, so it remains at the same height above the surface, apparently weightless.

G

Gagarin, Yuri (Alexeyevich) (1934–1968)

The first human being to travel in space, Gagarin was a Soviet cosmonaut who in 1961 was launched into **orbit** around the Earth aboard the spacecraft *Vostok 1*. Gagarin was born in the Smolensk region. He became a pilot in 1957, and on 12 April 1961 completed one orbit of the Earth, taking 108 minutes from launch to landing. He died in a plane crash while training for the *Soyuz 3* mission.

galactic coordinates

A system for measuring the position of astronomical objects on the **celestial sphere**. Positions are defined with reference to the galactic equator, the mid-plane of the disc of our **Galaxy**. Galactic coordinates are often used when astronomers are studying the distribution of material in the Galaxy.

LATITUDE AND LONGITUDE

- Galactic latitude is measured in degrees from the galactic equator (0°) to the north galactic pole (90°) and south galactic pole (–90°).
- Galactic longitude is measured in degrees eastward (0° to 360°) from a fixed point in the constellation of **Sagittarius** that approximates to the centre of the **Galaxy**.

---

galaxy

A system of millions or billions of stars, together with gas and **dust**, held together by **gravity**. Galaxies vary in size, structure, and luminosity, and, like stars, are found alone, in pairs, or in clusters. As these systems are very remote, they appear in telescopes as hazy, nebulous objects and were first described as **nebulae**. Later, when their remoteness was understood, they were known as 'island universes' or 'extra-galactic nebulae'.

Most galaxies occur in clusters, containing anything from a few to thousands of members. Our own galaxy is called the **Galaxy** (with a capital 'G') or the **Milky Way**.

Types of galaxy
Normal galaxies were classified by the US astronomer Edwin **Hubble** into three basic types: spiral, elliptical, and irregular.

Spiral galaxies, such as the Milky Way, are flattened in shape, with a central bulge of old stars surrounded by a disc of younger stars, arranged in spiral arms like a Catherine wheel. The arms of spiral galaxies contain gas and dust from which new stars are still forming. Spiral galaxies are classified according to the appearance of their arms:

- Sa spirals have a large nuclear bulge and tightly coiled spiral arms.
- Sc spirals have a small nucleus and arms less tightly wound.
- Sb spirals are intermediate between Sa and Sc.
- **Barred spirals** are classified from SBa to Sbc. They are spiral galaxies that have a straight bar of stars across their centre, from the ends of which the spiral arms emerge.

Elliptical galaxies are something like huge **globular clusters**, with no spiral arms. They are divided into eight subgroups, E0–E7, the E0s appearing spherical and the E7s the most elongated. Elliptical galaxies contain old stars and very little gas. They include the most massive galaxies known, containing a trillion stars. At least some elliptical galaxies are thought to be formed by the merging of spiral galaxies.

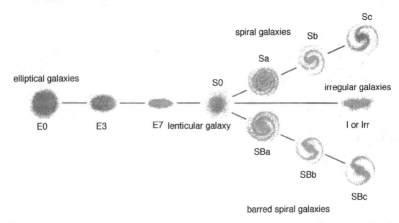

galaxy *Galaxies were classified by US astronomer Edwin Hubble in 1925. He placed the galaxies in a 'tuning-fork' pattern, in which the two prongs correspond to the barred and non-barred spiral galaxies.*

Irregular galaxies have a chaotic appearance and show no symmetry. Irregulars are very much less frequent than spirals and ellipticals; they also tend to be smaller but brighter in proportion to their mass. They contain large amounts of gas and dust, and large numbers of young, bright stars.

Starburst galaxies are spiral galaxies that appear unusually bright in the infrared part of the **spectrum** due to a recent burst of star formation. This may have been triggered by the gravitational influence of a companion galaxy, or even by a recent collision with such a galaxy.

MEASURING THE DISTANCES OF GALAXIES

- About 20 galaxies are known to be within 2.5 million **light years** away, and several thousand within 50 million light years.
- The distances of galaxies closer than 10 million light years can be estimated from the brightness of individual **Cepheid** variables, if such stars can be identified.
- Up to about 100 million light years away the **magnitude**s of **supergiants**, and of **novae** or **supernovae** at maximum, can be used to determine distance.
- Still greater distances have been estimated by comparing the apparent magnitude of a galaxy with its absolute magnitude.
- The greatest distances of all are found by measuring a galaxy's **red-shift** and assuming the truth of **Hubble's Law**.

The masses of galaxies

Once the distance of a galaxy is known, it becomes possible to estimate its mass, in the case of some of the nearer galaxies. It is also possible to estimate the masses of clusters of galaxies, but the masses so found have been larger than would be expected from the sum of the masses of the visible individual galaxies. This discrepancy, sometimes known as the problem of the **missing mass**, has not been explained.

Naked-eye galaxies

Only two galaxies, the **Magellanic Clouds** in the southern sky, are easily visible to the naked eye. The next brightest, the **Andromeda galaxy** in the northern sky, is just visible. About 35 of the brightest galaxies appear in the **Messier Catalogue**, the list compiled by the French astronomer Charles Messier (1730–1817). Several thousand appear in the New General

Catalogue. More than 100 million can be photographed with modern **telescopes**.

galaxy cluster

A grouping of **galaxies**, held together by the force of **gravity**. The vast majority of galaxies belong to clusters. Clusters of galaxies can be roughly classified as regular or irregular. Regular clusters have spherical symmetry, central concentration, and usually at least 1,000 members brighter than absolute **magnitude** –16. One of the nearest examples is in the constellation **Corona Borealis**. Irregular clusters are made up of loose groups of small clusters. Unlike the regulars, which consist almost entirely of ellipticals, the irregular clusters contain all types of galaxies. They vary greatly in content and may contain more than a thousand galaxies, as in the cluster in **Virgo**; or only 20 or so, as in the **Local Group** to which our **Galaxy** belongs.

Galaxy, the

The star-system to which our **Solar System** belongs is called the Galaxy (with a capital 'G' to distinguish it from other **galaxies**). It is often called the **Milky Way**. It is about 100,000 **light years** in diameter, and contains at least 100 billion stars. It is a member of a small cluster, the **Local Group**. The Solar System lies in one of its spiral arms, about 25,000 light-years from the centre.

The Galaxy

Diameter	100,000 light years
Distance of Sun from centre	30,000 light years
Number of stars	100 billion
Number of globular clusters	*c.* 150
Period of revolution of Sun around Galactic centre	225 million years

all values are approximate

Galileo (1564–1642)

Italian mathematician, astronomer, and physicist, whose full name was Galileo Galilei. He developed the astronomical **telescope** and was the first to see **sunspots**, the four main **satellites** of **Jupiter**, and the **phases** of the planet **Venus**. He clashed with the Church over the **Copernican system** of the universe, in which the Earth moves around the Sun, and was forced to deny his beliefs. He made enormous contributions to physics, helping to create the modern mathematical and experimental science.

Astronomical work

In July 1609, hearing that a Dutch scientist had made a telescope, Galileo worked out the principles involved and made a number of telescopes. He immediately made a series of spectacular discoveries:

- He discovered four satellites revolving around Jupiter, showing that the Earth is not the only centre of revolution in the **Solar System**. He compiled tables of their movements and proposed that they could be used as a 'celestial clock' to determine **longitude** on land and at sea.

- He observed the mountains of the Moon, showing that the Moon is a world like the Earth.

- He observed the phases of Venus, showing that one planet, at least, goes around the Sun.

Galileo Galilei, Italian mathematician, astronomer, and physicist.

- He saw myriad faint stars making up the **Milky Way**, showing that the universe hitherto observed with the naked eye was only a small part of the whole.

- He discovered **sunspots**, showing that the Sun is not a 'perfect' body, as suggested by ancient philosophers.

Galileo's results, published in *The Starry Messenger* (1610), were revolutionary. They made plausible for the first time the Sun-centred theory of the Polish astronomer Nicolaus **Copernicus**. Galileo continued to believe, however - following both Greek and medieval tradition – that orbits must be circular, not elliptical, in order to maintain the fabric of the cosmos in a state of perfection.

❝ Eppur si muove
And yet it does move. ❞

Traditionally supposed to have been whispered by **Galileo**
immediately after his recantation

GALILEO'S LIFE

1564 Galileo is born in Pisa, where he is educated.

1589 He becomes professor of mathematics at the University of Pisa.

1592 He accepts a position as professor at Padua, as his revolutionary discoveries have made him many enemies at Pisa. At Padua he writes a treatise on the specific gravities of solid bodies, and experiments with falling bodies.

1593 Galileo invents a thermometer, based on the expansion of air when it is warmed.

1609 He makes a telescope and uses it to observe the skies.

1610 He is appointed chief mathematician to the grand duke of Tuscany, Ferdinand II.

1616 Galileo's attempts to explain biblical texts in the light of his theories go against accepted opinion, and he is compelled by the Roman Inquisition not to assert 'what seems to contradict Scripture'.

1632 Galileo publishes *Dialogues on the Two Chief Systems of the World*, in which he promotes the theory of a Sun-centred Solar System. The book is banned by the church authorities in Rome.

1633 He is tried for heresy and forced to abjure his belief that the Earth moves around the Sun. He is to recite the seven Penitential Psalms once a week for three years, and is put under house arrest for his last years.

1638 Galileo publishes *Discourses and Mathematical Demonstrations Concerning Two New Sciences*.

1642 Galileo dies, still under house arrest, at Arcetri, near Florence.

❝ In my studies of astronomy and philosophy I hold this opinion about the universe, that the Sun remains fixed in the centre of the circle of heavenly bodies, without changing its place: and the Earth, turning upon itself, moves round the Sun. ❞

Galileo, letter to Cristina di Lorena, 1615

Galileo mission

Spacecraft that has explored **Jupiter's** system of **satellites** and also sent a probe into the planet's **atmosphere**. The mission was launched from the **space shuttle** *Atlantis* in October 1989, and took six years to journey to

Jupiter. *Galileo's* probe entered the atmosphere of Jupiter in December 1995. It radioed information back to the orbiter for 57 minutes before it was destroyed by atmospheric pressure.

- Despite technical problems data are still being relayed to Earth, but very slowly.

- In 1998 *Galileo* completed two flybys of Jupiter's icy moon **Europa**, but failed to collect any data because of technical errors.

- The spacecraft flew past **Venus** in February 1990 and passed within 970 km/600 mi of Earth in December 1990, using the gravitational fields of these two **planets** to increase its velocity.

- *Galileo* flew past the **asteroids** Gaspra in 1991 and Ida in 1993, taking close-up photographs.

- At the end of July 1995, while 55 million km/34 million mi from Jupiter, *Galileo* entered a dust storm and began detecting up to 20,000 particles a day (previously the maximum detected was 200).

- The **dust** is associated with Jupiter, and may come from the planet itself, the **planetary rings** surrounding it, or its **satellite**s.

gamma ray
The highest-energy form of electromagnetic radiation. Gamma rays generally have **wavelengths** shorter than those of **X-rays**, though the two terms are not precisely defined and there is some overlap. Gamma rays are produced on Earth in some forms of radioactivity, but they also reach the Earth from space in **cosmic radiation**.

gamma-ray astronomy
The study of **gamma rays** from space. This high-energy radiation does not reach the ground and so ground-level instruments must detect them indirectly by observing the showers of particles produced when a gamma ray is absorbed by atoms in the **atmosphere**. Gamma-ray detectors can be carried above the atmosphere on spacecraft; they are very different from optical **telescopes** because gamma rays cannot be focused as **light** can. Much of the gamma radiation detected comes from collisions between **hydrogen** gas and **cosmic rays** in our **Galaxy**. Some sources have been identified, including the **Crab Nebula** and the Vela **pulsar** (the most powerful gamma-ray source detected).

See also: *gamma-ray burster.*

IN SEARCH OF GAMMA RAYS

▪ The first gamma-ray satellites were SAS II (1972) and COS B (1975). SAS II failed after only a few months, but COS B continued working until 1982, carrying out a complete survey of the galactic disc.

▪ The Compton Gamma Ray Observatory was launched by the US space shuttle *Atlantis* in April 1991 to study the gamma-ray sky for five years. The observatory cost $617 million and at 15 tonnes/17 tons was the heaviest payload ever carried by a space shuttle.

---

gamma-ray burster

Mysterious events in which enormous quantities of **gamma radiation** are given out in a few seconds or a few minutes. They were discovered by Vela **satellites** in the early 1970s, and scores of them are observed each year. They occur at immense distances, billions of **light years** away and therefore billions of years in the past. In May 1999 a gamma-ray burst was seen that occurred 7 billion **light years** away, when the universe was half its present age. In 100 seconds it released as much energy as our **Galaxy** emits in the visible part of the spectrum in 30 years. An even more distant one had been observed in 1998, over 8.3 billion light years away. Cosmologists have suggested **black holes** merging with each other or with **neutron** stars to account for such huge explosions.

See also: *gamma-ray astronomy.*

Ganymede

A moon of the planet **Jupiter**, and the largest satellite in the **Solar System** (larger than the planet **Mercury**). Its surface is a mixture of cratered and grooved terrain. Molecular oxygen was identified on Ganymede's surface in 1994.

The *Galileo* **mission** detected a **magnetic field** around Ganymede in 1996; this suggests it may have a molten core. *Galileo* photographed Ganymede at a distance of 7,448 km/4,628 mi. The resulting images were 17 times clearer than those taken by *Voyager* in 1979, and showed the surface to be extensively cratered and ridged, probably as a result of forces similar to those that create mountains on Earth. In March 1997 *Galileo* also detected molecules containing both carbon and nitrogen on the surface. Their presence may indicate that Ganymede harboured life at some time.

GANYMEDE: STATISTICS

Diameter	Distance from centre of planet	Period
5,260 km/3,270 mi	1.1 million km/700,000 mi	7.2 days

gegenschein

A pale glow that can sometimes be seen on a very dark night at the point in the sky that is directly opposite to the Sun. It is caused by sunlight shining on the tiny particles that make up **interplanetary matter**.

Gemini

A prominent **constellation** in the **zodiac** in the northern hemisphere. **Pollux** is its brightest star; its second-brightest is **Castor**, a system of six stars.

- The Sun passes through Gemini from late June to late July.

- Each December, the Geminid **meteors** radiate from Gemini.

- In astrology, the dates for Gemini are between about 21 May and 21 June (see **precession**).

- In Greek mythology, Castor and Pollux were the twin sons of Leda and Zeus, king of the gods, and brothers of Helen of Troy. They were protectors of mariners, and were transformed at their death into the constellation Gemini.

Gemini missions

A series of US space flights (1965–66) in which astronauts practised rendezvous and docking of spacecraft, and working outside them, in preparation for the *Apollo* **missions** to the Moon. *Gemini* spacecraft carried two astronauts and were launched by *Titan* rockets.

Gemini telescopes

An international project to build a pair of giant astronomical **telescopes** on high mountain sites in Hawaii and Chile.

The chosen sites, Mauna Kea, Hawaii (4,200 m/13,800 ft high) and Cerro Pachón, Chile (2,700 m/8,860 ft), have excellent observing conditions and

---

GEMINI'S MIRRORS

▪ Each telescope has a primary mirror 8 m/26 ft in diameter and 20 cm/8 in thick, constructed of ultra-low-expansion glass.

▪ The shape of each mirror is continually adjusted by 120 active supports. These prevent the mirror sagging as the telescope moves across the sky.

▪ The supports also compensate for the distorting effects of the **atmosphere**, producing sharp images throughout the visual and near-infrared parts of the **spectrum**. This technique is called adaptive optics.

between them cover the whole sky. The $184 million project is funded by the USA, UK, Canada, Chile, Argentina, and Brazil. The Hawaiian instrument, Gemini North, was inaugurated in June 1999, while the Chilean instrument was still under construction.

geocentric theory
Any astronomical theory that states that the Earth is at the centre of the universe, with the Sun and (generally) all other celestial bodies revolving around it. With the exception of some bold Greeks such as **Aristarchus of Samos**, this picture was assumed almost without question by most thinkers until the 16th century. Then Nicolaus **Copernicus** published his **heliocentric theory** (Sun-centred theory), and it gradually won acceptance. An unusual variant of a geocentric theory was that of Tycho **Brahe**, who proposed that the planets go round the Sun while the Sun goes around the Earth.

geostationary orbit
A circular path 35,900 km/22,300 mi above the Earth's equator. A **satellite** moving in this **orbit** takes 23 hours, 56 minutes and 4.1 seconds, moving from west to east, to complete an orbit. Since this is the precise time that it takes the Earth to revolve once, the satellite appears to hang stationary over one place on the Earth's surface. Geostationary orbits are used particularly for communications satellites and weather satellites.

ORIGINS AND ORBITS

- The idea of communications satellites in geostationary orbits was first thought of by the British science fiction author Arthur C Clarke.
- A *geosynchronous* orbit lies at the same distance from Earth but is inclined to the equator.

giant star
Huge, bright stars that have swollen up towards the ends of their lives. Giants have exhausted their supply of **hydrogen** fuel and derive their energy from the fusion of **helium** and heavier elements. They are roughly 10–300 times bigger than the Sun, with 30–1,000 times the luminosity. The cooler giants are known as **red giant**s. The largest are called **supergiants**. Giant stars lie to the top right of the **Hertzsprung–Russell diagram**.

Giotto mission
Space probe built by the **European Space Agency** to study Halley's **comet**.

- Launched by an *Ariane* rocket in July 1985, *Giotto* passed within 600 km/375 mi of the comet's nucleus on 13 March 1986.

- At its closest approach, *Giotto* passed the comet at a relative speed 50 times greater than that of a bullet.

- *Giotto* was sandblasted by a hail of fine particles thrown off by the comet. Its camera was blacked out moments before closest approach, but not before sending back spectacular images of the cratered nucleus

- On 2 July 1990, it flew past Earth at a distance of 23,000 km/14,000 mi, which diverted its path to encounter another comet, Greg-Skjellerup, on 10 July 1992.

- The probe was named after the Italian painter Giotto (*c.* 1267–1337).

Glenn, John Herschel, Jr (1921–   )

The first US astronaut to **orbit** the Earth. On 20 February 1962, John Glenn circled the Earth three times in the **Mercury** spacecraft *Friendship 7*, in a flight lasting 4 hr 55 min.

LIFE AFTER SPACE

- After retiring from NASA, Glenn was elected to the US Senate as a Democrat for Ohio in 1974. He was subsequently re-elected in 1980 and 1986.

- Glenn became the oldest person in space when, at the age of 77, he embarked on a nine-day mission aboard the shuttle Discovery in October 1998.

- As a senator, he advocated nuclear-arms-production limitations and increased aid to education and job-skills programmes.

- He unsuccessfully sought the Democratic presidential nomination in 1984.

globular cluster

A spherical or near-spherical star cluster containing approximately 10,000 to millions of stars. More than a hundred globular clusters are distributed in a spherical halo around our **Galaxy**. They consist of old stars, formed early in the Galaxy's history. Globular clusters are also found around other **galaxies**.

granulation, solar

A mottling of the **photosphere**, the bright visible surface of the Sun. The 'granules' are cells of rising hot and bright gas, interspersed with cells of gas that has cooled, is less bright, and is sinking. Each granule is about 1,000 km/600 mi across.

gravitational constant

The number that describes the strength of **gravity**. It is one of the funda-mental physical constants that physicists would like to explain. Its value is 6.67×10^{-11} m^3 kg^{-1} s^{-2}.

gravitational lensing

The bending of **light** by a gravitational field, predicted by **Einstein's** general theory of **relativity**. The effect was first detected in 1917 when the light from stars was found to be bent as it passed the totally eclipsed Sun. More remarkable is the splitting of light from distant **quasars** into two or more images by intervening **galaxies**. In 1979 the first double image of a quasar produced by gravitational lensing was discovered and a quadruple image of another quasar was later found.

gravitational wave

A disturbance in **space-time** that spreads out at the speed of light from a strongly accelerating massive object, such as binary **pulsars** – pulsars in **orbit** around other objects. Their existence was predicted in Albert **Einstein's** general theory of **relativity**, but they have not yet been directly observed. However, some binary pulsars have been found to lose energy at just the rate expected if they are radiating gravitational waves. Huge detec-tors are being built on Earth and others are planned to go into space in the early part of the 21st century.

See also: *graviton.*

graviton

A hypothetical particle that, if it exists, is the counterpart of the **gravitational wave**. According to quantum theory, all fields (such as **gravity**) are associ-ated with particles – for example, **light** and other radiations consist of electromagnetic fields, and also of particles called photons. Physicists are searching for gravitational waves and gravitons.

gravity

A force of attraction that exists between any two objects in the universe, arising from their masses. It is regarded as one of the four fundamental forces of nature, the other three being the electromagnetic force, the strong nuclear force, and the weak nuclear force. The gravitational force is the weakest of the four forces, but it acts over great distances, whereas the two nuclear forces are extremely short-range – of the order of the size of the atomic nucleus. Electromagnetism is also long-range, but the existence of equal amounts of negative and positive charge in the universe makes it less dominant than gravity, which is always attractive and therefore cumulative. Gravity therefore dominates the universe on planetary scales and larger. The particle believed to be the carrier of the gravitational force is the **graviton**.

AN ATTRACTIVE MOUNTAIN

One of the earliest gravitational experiments was undertaken by Nevil Maskelyne in 1774. He measured the force with which the Scottish Mount Schiehallion attracted a plumb-bob.

Newton's law of gravitation
Isaac **Newton's** law of universal gravitation states that the force between two objects is proportional to their masses (that is, if either of the masses is doubled, the gravitational force between them is doubled, and so on); and inversely proportional to the square of the distance between them (that is, if the distance between them is doubled, the force falls to a quarter; if it is tripled, it falls to a ninth, and so on). In symbols this is:

$$F = \frac{Gm_1 m_2}{r^2}$$

Here, G is a constant called the **gravitational constant**. If the objects are, say, the Earth and an apple, this equation gives both the force exerted on the apple by the Earth, and that exerted on the Earth by the apple. They are equal, though opposite – an instance of Newton's Third Law of Motion, that for every action (force) there is an equal and opposite reaction. The gravitational force on the small apple has a large effect – the apple falls. The equal and opposite force on the huge Earth has a tiny and quite unnoticeable effect.

Gravity and relativity
Albert **Einstein's** general theory of **relativity** treats gravitation not as a force but as the curvature of **space-time** around a body. Relativity predicts that a gravitational field bends **light** and **red-shifts** it (lengthens its wavelength). Both effects have been observed.

Another prediction of relativity is the existence of **gravitational waves**, which should be produced when massive bodies are violently disturbed. These waves are so weak that they have not yet been detected with certainty.

IT'S NOT A GAS

Astronauts in space cannot burp. It is gravity that causes bubbles to rise to the top of a liquid, so **space shuttle** crews were forced to request less gas in their fizzy drinks to avoid discomfort.

gravity assist
Using the **gravity** of a **planet** or **satellite** to boost the speed of a spacecraft, enabling it to travel farther on a given amount of fuel. The *Galileo* **mission**, for example, made repeated fly-bys of **Venus** and the Earth to build up sufficient speed to be flung outwards towards its destination, **Jupiter**.

great circle
A circle on the Earth or the **celestial sphere** whose centre is the centre of the Earth. On Earth the equator is an example. On the celestial sphere the celestial equator is an example of a great circle, and the **ecliptic** is another.

Green Bank Telescope
The largest fully steerable **radio telescope** in the world. It is located at Green Bank, Pocahontas County, West Virginia, in the United States, home of the National Radio Astronomy Observatory (NRAO). Its dish has a diameter of 110 m/361 ft. It replaces a 91-m/300-ft dish that collapsed in November 1988.

greenhouse effect
The warming of a **planet's** surface by the trapping of solar energy by its **atmosphere** (if the planet has one). The natural greenhouse effect is responsible for making the Earth a habitable planet. **Mars** is too cold because of a lack of sufficient quantities of greenhouse gases. **Venus** is scorchingly hot because of an excess of such gases. On Earth there may be an additional 'anthropogenic' or human-created greenhouse effect, owing to the increased quantities of carbon dioxide, methane, and other gases put into the atmosphere by industry and agriculture. The United Nations Environment Programme estimates that by 2025, average world temperatures could have risen by 1.5°C/2.7°F with a consequent sea-level rise of 20 cm/7.9 in. Low-lying areas – even entire countries – would be threatened by flooding and crops would be affected by the change in climate. However, predictions about global warming and its possible climatic effects are tentative and often conflict with each other.

How the greenhouse works

- Solar radiation is absorbed by the Earth and re-emitted from the surface as longer-wave **infrared radiation**. This is prevented from escaping by various gases in the air.
- Greenhouse gases trap heat because they readily absorb infrared radiation.
- The main greenhouse gases are carbon dioxide, methane, and chlorofluorocarbons (CFCs), as well as water vapour.
- Fossil-fuel consumption and forest fires are the principal causes of carbon dioxide build-up; methane is a by-product of agriculture (rice, cattle, sheep).

Naming the greenhouse
The greenhouse effect was named after the Swedish scientist Svante Arrhenius (1859–1927). But it was first predicted in 1827 by the French mathematician Joseph Fourier (1768–1830).

STARTLING FACTS

- The concentration of carbon dioxide in the atmosphere is estimated to have risen by 25% since the Industrial Revolution, and by 10% since 1950. The rate of increase is now 0.5% a year.
- Chlorofluorocarbon (CFC) levels are rising by 5% a year, and nitrous oxide levels by 0.4% a year.
- Arctic ice was 6–7 m/20–23 ft thick in 1976, but had decreased to 4–5 m/13–17 ft by 1987.
- A medium-sized power station, generating around 500 megawatts, produces 500 tonnes of carbon in the form of carbon dioxide, every hour.
- Aircraft vapour trails could be contributing to global warming. They trap heat, increasing the warming of the atmosphere beneath them. German research indicates that a tenth of the cirrus cloud over central Europe is produced by aircraft.

Greenwich Mean Time (GMT)
Mean **solar time** on the zero line of **longitude** (the Greenwich **meridian**), which passes through the old **Royal Greenwich Observatory** in London. Mean solar time is time as defined by the **mean** Sun at Greenwich. In scientific work GMT was replaced in 1986 by coordinated universal time (UTC), but continued to be used to measure longitudes and define the world's standard time zones.
   See also: *time measurement.*

Halley, Edmond (1656–1742)

English astronomer most famous for identifying the **comet** that was later to be known by his name. He also compiled a star catalogue, detected the **proper motion** of stars using historical records, and began a line of research that, after his death, resulted in a reasonably accurate calculation of the **astronomical unit**.

Halley made many other notable contributions to astronomy, including the discovery of the **proper motions** of **Aldebaran**, Arcturus, and **Sirius**, and working out a method of obtaining the solar **parallax** by observations made during a transit of **Venus**. He was also a pioneer geophysicist and meteorologist, and worked in many other fields, including mathematics and archaeology. He was a friend of the scientist Isaac **Newton**, and it was at Halley's urging and his expense that Newton's great work the *Principia* was published.

In 1698–1700, Halley captained a small vessel that he used to observe magnetic variation (the deviation of the compass from true north) over the Atlantic Ocean from 52° north to 52° south, in the hope that he might be able to use the information as a means of determining longitude.

PENETRATING THE DISGUISE

Halley calculated that comet sightings reported in 1456, 1531, 1607, and 1682 were all of the same object. Until then it had been assumed that different appearances of comets represented distinct objects. He predicted that the comet would reappear in 1758. When it did so, after his death, public acclaim for the astronomer was such that his name was irrevocably attached to it.

---

halo

1 The gas around the nucleus of a **comet**.
2 The spherical collection of **globular clusters** forming a 'shell' surround-

ing our otherwise compact, disc-shaped **Galaxy**.

3 A bright disc or circle of **light** around the Sun or Moon, caused by ice crystals in the Earth's **atmosphere**.

Hawking, Stephen (William) (1942–)

British physicist whose work in **general relativity** has contributed to the search for a quantum theory of **gravity** to explain **black holes** and the **Big Bang**. His book *A Brief History of Time* (1988) gives a popular account of cosmology and became an international best-seller, making Hawking the most famous scientist since Albert **Einstein**. He is almost totally paralysed by motor neurone disease, and uses a wheelchair and a speech synthesizer.

In 1974 Hawking published perhaps his most remarkable result: that **black holes** are not entirely black – they can emit particles in the form of thermal radiation. This is now called Hawking radiation.

Out of the blackness

Hawking has proposed a physical explanation for Hawking radiation which relies on the quantum-mechanical concept of 'virtual particles'. These exist as particle–antiparticle pairs and fill 'empty' space. Hawking suggested that, when such a pair is created near a black hole, one member of the pair might disappear into the black hole, leaving the other particle, which could escape to infinity. This would be seen by a distant observer as thermal radiation.

heliocentric theory

The theory that the Sun lies at the centre of the **Solar System**, and therefore that the Earth is a **planet** like the others. **Aristarchus of Samos** may have been the first to hold this theory, but it was first seriously considered when Nicolaus **Copernicus** proposed it at the end of his life, in 1543.

See also: *geocentric theory.*

heliosphere

The huge region around the **Solar System** where the **solar wind** dominates **interstellar matter**. The boundary of this region is called the *heliopause* and is believed to lie about 100 **astronomical units** from the Sun, where the flow of the solar wind merges with the interstellar gas.

helium (Greek *helios* 'Sun')

The second most abundant element (after **hydrogen**) in the universe. Under normal conditions on Earth it is a colourless, odourless, gas. It is present in small quantities in the Earth's **atmosphere**, produced by alpha decay of radioactive elements in the Earth's crust.

- Helium is the second lightest element after hydrogen.

- Helium is a component of most stars, including the Sun, where the nuclear-fusion process converts hydrogen into helium with the production of heat and light.

- Some helium was present in the universe immediately after the **Big Bang**; the rest has been 'cooked' since then in the cores of stars.

- Helium was discovered when spectral lines of an unknown element were observed in the Sun's **spectrum**. It was given its name from the Greek word *helios*, meaning 'Sun'.

Hercules
The fifth-largest **constellation**, lying in the northern hemisphere of the sky. Despite its size it contains no prominent stars. Its most important feature is M13, the best example in the northern hemisphere of a **globular cluster**. M13 is 22,500 **light years** from Earth, and from Earth seems to lie between Eta and Zeta Herculis.

Hermes (vehicle)
A spaceplane proposed by the **European Space Agency** to ferry astronauts to and from the **International Space Station**. *Hermes* would have been launched on an *Ariane* rocket and flown back to Earth to land on a conventional runway. The project was abandoned in the early 1990s because of financial pressures.

Herschel, Caroline (Lucretia) (1750–1848)
German-born English astronomer, sister of William **Herschel**, and from 1772 his assistant.

Between 1786 and 1797 Caroline Herschel discovers eight comets, five undoubtedly unobserved before, and many of the smaller nebulae and star clusters included in her brother's catalogue. The Royal Astronomical Society awards her their gold medal in 1828, and, in 1835, make her an honorary member.

Herschel, (Frederick) William (1738–1822)
German-born English astronomer who was one of the greatest of observers. He was a skilled **telescope**-maker, and pioneered the study of **binary stars**

> **❝**I have looked farther into space than ever a human being did before me. **❞**
>
> **William Herschel**

Herschel *William Herschel holding a diagram of Uranus and its satellites.*

and **nebulae**. He discovered the planet **Uranus** in 1781 and **infrared radiation** in sunlight. He catalogued over 800 binary stars, and found over 2,500 nebulae, which were catalogued by his sister Caroline **Herschel**; this work was continued by his son John **Herschel**. By studying the distribution of stars, William established the basic form of our **Galaxy**, the **Milky Way**.

TELESCOPES AND PRISMS

- In 1879, Herschel builds, in Slough, west London, a 1.2-m/4-ft telescope of 12 m/40 ft focal length, the largest in the world at the time. But he makes most use of a more satisfactory 46-cm/18-in instrument.
- In 1800 Herschel examines the solar spectrum using prisms and temperature-measuring equipment, and finds that the hottest radiation is infrared.

Herschel, John (Frederick William) (1792–1871)

English scientist, astronomer, and photographer who discovered thousands of close **binary stars**, clusters, and **nebulae**. He coined the terms 'photography', 'negative', and 'positive', discovered sodium thiosulphite as a fixer of silver halides, and invented the cyanotype process; his inventions also include astronomical instruments. His *General Catalogue of 10,300 Multiple and Double Stars* was published posthumously. His *General Catalogue of Nebulae and Clusters* coordinated into one catalogue the results of surveys made by him, by his father William **Herschel**, and by other astronomers.

Hertzsprung–Russell diagram

A graph on which the surface temperatures of stars are plotted against their **luminosities** (their true brightnesses, after correcting for their differing distances from us). It is of fundamental importance in describing stars and understanding their life-histories. Most stars, including the Sun, fall into a narrow band called the main sequence. When a star grows old it moves from the main sequence to the upper right part of the graph, into the area of the giants and **supergiants**. At the end of its life, as the star shrinks to become a **white dwarf**, it moves again, to the bottom left area. It is named after the Danish astronomer Ejnar Hertzsprung (1873–1967) and the US astronomer Henry Norris Russell (1877–1957), who independently devised it in the years 1911–13.

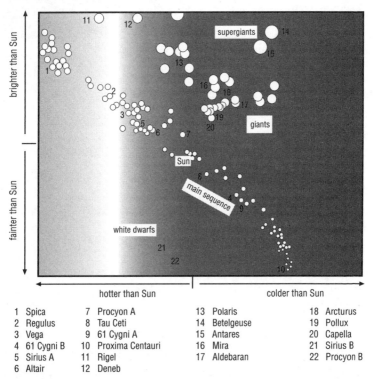

1	Spica	7	Procyon A	13	Polaris	18	Arcturus
2	Regulus	8	Tau Ceti	14	Betelgeuse	19	Pollux
3	Vega	9	61 Cygni A	15	Antares	20	Capella
4	61 Cygni B	10	Proxima Centauri	16	Mira	21	Sirius B
5	Sirius A	11	Rigel	17	Aldebaran	22	Procyon B
6	Altair	12	Deneb				

Hertzsprung-Russell diagram *The Hertzsprung–Russell diagram relates the brightness (or luminosity) of a star to its temperature. Most stars fall within a narrow diagonal band called the main sequence. A star moves off the main sequence when it grows old. The Hertzsprung–Russell diagram is one of the most important diagrams in astrophysics.*

Hipparchus (*c.* 190 BC – *c.* 120 BC)

Greek astronomer and mathematician who calculated the lengths of the **solar year** and the lunar month. He was born in Nicaea, Bithynia (now in Turkey), and lived on the island of Rhodes and in Alexandria, Egypt.

> ❝ The difference in the length of the year can be accurately observed from the records on the bronze sphere in the Square Stoa at Alexandria. ❞
>
> **Hipparchus** *On the Length of the Year*

ACHIEVEMENTS OF HIPPARCHUS

- In 134 BC Hipparchus noticed a new star in the constellation **Scorpio**, a discovery which inspired him to compile the first star catalogue.
- He entered his observations of stellar positions using a system of celestial latitude and longitude, and taking the precaution wherever possible of stating the alignments of other stars as a check on present position.
- He advanced **Eratosthenes'** method of determining the situation of places on the Earth's surface by lines of **latitude** and **longitude**.
- His catalogue contained 850 stars, which he classified by **magnitude** (brightness).
- His accurate observations enabled him to discover the **precession** of the **equinoxes.**
- His catalogue, completed in 129 BC, was used by Edmund **Halley** some 1,800 years later.

Hipparcos

An astronomical **satellite** launched by the **European Space Agency** in 1989. It is named after the Greek astronomer **Hipparchus** – or alternatively, its name can be regarded as an acronym for *Hi*gh *Pr*ecision *Pa*rallax *Co*llecting *S*atellite). It is the world's first astrometry ('star-position measuring') satellite and provides precise positions, distances, colours, brightnesses, and apparent motions for over 100,000 stars.

hour-angle
The angle between a celestial body and the observer's **meridian**. It is expressed in hours, minutes, and seconds of time and is equal to the time until the body passes due south or due north of the observer, or the time that has elapsed since it did so.

hour-circle
An imaginary circle drawn on the **celestial sphere** and passing through the celestial poles. The zero hour-circle is the one passing over the observer, the **meridian**.
   See also: *hour-angle.*

Hoyle, Fred(erick) (1915–   )
British astronomer, cosmologist, and writer whose astronomical research has dealt mainly with the internal structure and evolution of the stars. Hoyle is most famous for having developed, with Hermann Bondi (1919–   ) and Thomas Gold (1920–   ), the **steady-state theory** of the universe as an alternative to the **Big Bang** theory.

❝ Space isn't remote at all. It's only an hour's drive away if your car could go straight upwards. ❞

Fred Hoyle, *Observer*, September 1979

Hubble, Edwin (Powell) (1889–1953)
US astronomer who discovered the existence of **galaxies** outside our own, classified them and showed that the universe is expanding. Nearly all Hubble's work related to '**nebulae**', which when he began his career referred both to nebulae in the modern sense, namely, interstellar gas clouds, and to objects that he was the first to show were other galaxies. It has been said that Hubble opened up the universe of galaxies in the same way that the astronomer and physicist **Galileo** opened up the **Solar System** in the 17th century, and the astronomers William **Herschel** and John **Herschel** opened up the **Milky Way** in the 18th and 19th centuries.
   In 1924 Hubble discovered the Cepheid variable stars in the Andromeda Nebula, proving it to be a galaxy for beyond our own, and a year later introduced the classification of galaxies as spirals, barred spirals, and ellipticals. (*See illustration on p. 66.*)
   See also: *Hubble's law.*

Hubble's law
The relationship between a **galaxy's** distance from us and the speed at which it is travelling away from us as the universe expands. It was

announced in 1929 by the US astronomer Edwin **Hubble**. Hubble found that galaxies are moving apart at speeds that increase in direct proportion to their distance apart. The rate of expansion is known as Hubble's constant. Observations suggest that galaxies are moving apart at a rate of 50–100 kps/30–60 mps for every million parsecs of distance. This means that the universe, which began at one point according to the **Big Bang** theory, is between 10 billion and 20 billion years old.

Hubble Space Telescope (HST)

An unmanned space-based astronomical **observatory**, orbiting the Earth at an altitude of 610 km/380 mi. The Hubble Space Telescope has sent back to Earth a wealth of spectacular views of the universe since it was launched in 1990. HST allows astronomers to observe the birth of stars, find **planets** around neighbouring stars, follow the expanding remnants of exploding stars, and search for **black holes** in the centre of **galaxies**. It is a cooperative programme between the **European Space Agency** (ESA) and the US space agency **NASA** (National Aeronautics and Space Administration), and is the first spacecraft specifically designed to be serviced in **orbit** as a permanent space-based observatory.

HUBBLE FACTS

▪ HST consists of a 2.4-m/94-in **telescope** and four complementary scientific instruments. It is roughly cylindrical, 13 m/43 ft long, and 4 m/13 ft in diameter, with two large winglike solar panels.

▪ Its instruments include a wide field/planetary camera designed to gather the sharpest astronomical images ever.

▪ The Faint Object Camera was built by ESA and uses an image intensifier to image the faintest object resolvable by the telescope.

▪ The Faint Object Spectrometer (FOS) measures spectra in a wide range of light, from ultraviolet to near-infrared.

▪ The Goddard High-Resolution Spectrograph is similar to the FOS but is dedicated to **ultraviolet astronomy** and **infrared astronomy**.

The benefits of space

By having a large telescope above Earth's **atmosphere**, astronomers are able to look at the universe with unprecedented clarity. Celestial observations by HST are unhampered by clouds and other atmospheric phenomena that distort and dim starlight. In particular, the apparent twinkling of starlight caused by density fluctuations in the atmosphere limits the clarity of ground-based telescopes. HST performs at least ten times better than such telescopes and can see almost to the edge of the universe and almost back to the beginning of time (see **Big Bang**).

Hubble at work

- In December 1995 HST was trained on an 'empty' area of sky near the Plough, now termed the Hubble Deep Field. Around 1,500 galaxies, mostly new discoveries, were photographed.

- Two new instruments were added in February 1997. The Near Infrared Camera and Multi-Object Spectrometer (NICMOS) enable Hubble to see things even farther away (and therefore older) than ever before. The Space Telescope Imaging Spectrograph will work 30 times faster than its predecessor, as it can gather information about different stars at the same time.

- Three new cameras had to be fitted shortly afterwards as one of the original ones was found to be faulty.

- In May 1997, three months after astronauts installed new equipment, US scientists reported that Hubble had made an extraordinary finding. Within 20 minutes of searching, it discovered evidence of a black hole 300 million times the mass of the Sun. It is located in the middle of galaxy M84 about 50 million **light years** from Earth.

- Further findings in December 1997 concerned the shapes of dying stars. Previously, astronomers had thought that most stars die surrounded by a globe of hot gas expanding into space. The photographs taken by the HST show other shapes, such as pinwheels. This may be indicative of how the Sun will die.

- Galaxies photographed by the HST in 1998 were born at least 12 billion years ago. These are the most remote objects ever imaged.

Huygens, Christiaan (1629–1695)
Dutch mathematical physicist and astronomer who proposed the wave theory of light, developed the pendulum clock in 1657, discovered the polarization of light, and observed **Saturn's** rings. His work in astronomy was an impressive defence of the **Copernican system**. Huygens was born in The Hague and studied at Leiden and Breda. He soon showed a strong talent for mathematics, and his future greatness was predicted by the French philosopher, mathematician and scientist Descartes. At various times Huygens lived in Denmark, Holland, France, and England.

- In 1657, Huygens developed a clock regulated by a pendulum, an idea that he published and patented. By 1658, major towns in Holland had pendulum tower clocks. He derived the relationship between the period of a simple pendulum and its length.

- The *Traité de la Lumière/Treatise on Light* of 1678 contained Huygens' wave or pulse theory of light. He theorized that light is transmitted as a pulse moving through a medium, or ether, by setting up a train of vibrations in the ether. His publication was partly a counter to Isaac **Newton's** particle theory of light.

- Huygens' comprehensive study of geometric optics led to the invention of a telescope eyepiece that reduced **chromatic aberration**. It consisted of two thin plano-convex lenses, rather than one thick lens, with the field lens having a focal length three times greater than that of the eyepiece lens. Its main disadvantage was that cross-wires could not be fitted to measure the size of an image. Huygens then developed a micrometer to measure the angular diameter of celestial objects.

THE RINGS OF SATURN

With a home-made **telescope** Huygens discovered **Titan**, one of Saturn's moons, in 1655. Later that year he observed that Titan's period of revolution was about 16 days and that it moved in the same plane as the so-called 'arms' of Saturn. This strange appearance of the planet phenomenon had been an enigma to many earlier astronomers, but because of Huygens' superior telescope, 7 m/23 ft long, he partially solved the mystery. In 1659, he published a Latin anagram that, when interpreted, read 'It (Saturn) is surrounded by a thin flat ring, nowhere touching, and inclined to the ecliptic'. The theory behind Huygens' hypothesis followed later in *Systema Saturnium* (1659), which included observations on the **planets**, their **satellites**, the **Orion Nebula** and a determination of the **period** ('year') of **Mars**, and provided further evidence for the Copernican view of the **Solar System**.

---

hydrogen

The most abundant element in the universe, accounting for 93% of the total number of atoms and 76% of the total mass. In normal conditions on the Earth, it is a colourless, odourless, gaseous, nonmetallic element, symbol H, atomic number 1, relative atomic mass 1.00797.

- Hydrogen is the lightest of all the elements and occurs on Earth chiefly in combination with oxygen as water.

- It is a component of most stars, including the Sun, whose heat and light are produced through the nuclear-fusion process that converts hydrogen into **helium**.

- When subjected to a pressure 500,000 times greater than that of the Earth's **atmosphere**, hydrogen becomes a solid with metallic properties, as in the interior of **Jupiter**.

Icarus

An **asteroid** 1.5 km/1 mi in diameter, discovered in 1949. It was the first asteroid known to approach the Sun closer than does the planet **Mercury**. It orbits the Sun every 409 days at a distance of 28–300 million km/18–186 million mi (0.19–2.0 **astronomical units**). In 1968 it passed 6 million km/ 4 million mi from the Earth. It is an **Apollo asteroid.**

impact crater

The bowl-shaped depression left in the surface of a **planet**, a moon, or the Earth by the fall of a **meteoroid**. Impact craters cover the surfaces of the **Solar System's** rocky airless worlds. On planets with geologically active surfaces or atmospheres, or both, such as the Earth and **Venus**, most of the traces of impact craters have been removed over time. Most impact cratering occurred billions of years ago, soon after the birth of the Solar System, as the newly formed planets and **satellites** swept up **interplanetary matter**.

See also: *crater.*

inferior planet

Either of the planets **Mercury** or **Venus**, whose orbits lie within that of the Earth. They are best observed when at their greatest **elongation** from the Sun, either at eastern elongation in the evening (setting after the Sun) or at western elongation in the morning (rising before the Sun).

inflation

A period of superfast expansion of the universe that occurred a minute fraction of a second after the birth of the universe in the **Big Bang**. The universe increased in size by an enormous ratio – from being smaller than an atom to the size of a grapefruit. This represented not an expansion of the universe into empty space but the creation of space itself. Before and after the inflation period, the universe was a seething mass of subatomic particles and radiation. Expansion continued afterwards at a slower rate, the universe cooled, matter built up into atoms, and **galaxies**, stars, and **planets** formed.

See also: *universe, future of.*

infrared astronomy
The study of **infrared radiation** produced by relatively cool gas and dust in space, as in the areas around newly forming stars. In 1983, **IRAS** (the Infrared Astronomy Satellite) surveyed the entire sky at infrared wavelengths. It found five new **comets**, thousands of **galaxies** undergoing bursts of star formation, and **planetary systems** forming around several dozen stars.

Planets and gas clouds emit their light in the far and mid-infrared region of the **spectrum**. The *Infrared Space Observatory* (ISO), launched in 1995, observes a broad wavelength band (3–200 micrometres) in this region. It is 10,000 times more sensitive than *IRAS*, and searches for **brown dwarfs** (cool gaseous bodies less massive than the Sun).

INFRARED TELESCOPES

An infrared **telescope** is designed to receive electromagnetic waves in the infrared part of the **spectrum**. Infrared telescopes are always reflectors (glass lenses are opaque to infrared waves) and are normally of the **Cassegrain** type. Since all objects at normal temperatures emit strongly in the infrared, careful design is required to ensure that the weak signals from the sky are not swamped by radiation from the telescope itself. Infrared telescopes are sited at high mountain **observatories** above the obscuring effects of water vapour in the **atmosphere**. Modern large telescopes are often designed to work equally well in both visible and infrared light.

---

infrared radiation
Invisible electromagnetic radiation lying just beyond the red end of the visible **spectrum**. It was discovered by the English astronomer William **Herschel**, and is often called 'heat radiation' because it is given out by objects at the temperatures of our normal surroundings. Infrared radiation coming from objects in space is largely blocked out by the Earth's **atmosphere**, but can be studied from mountain-tops and from spacecraft. It is defined as occupying the wavelength range from about 0.7 micrometre (at the wavelength limit of red light) to about 1 millimetre.
   See also: *infrared astronomy, ultraviolet radiation.*

interferometry
Any of several techniques used in astronomy to obtain high-resolution images of astronomical objects. It involves interference between two rays of

light or beams of radio waves from celestial objects. Interference results in the two sets of waves increasing in intensity or cancelling each other out, depending on whether the waves are 'in step' or 'out of step'. Interferometry has long been applied in the laboratory to make extremely accurate measurements of distance. **Radio telescopes** can be linked for purposes of interferometry, enabling them to distinguish far more detail. The same technique is now being applied in optical telescopes, to make measurements of tiny details such as **sunspots** on stars.

International Space Station (ISS)
A 430-tonne orbiting **space station** being constructed by the United States, Russia, Japan, Canada, and the **European Space Agency**. The complete station will measure 88 m/288 ft in length and 108 m/354 ft across its solar wings. Up to seven astronauts will live and work in a space equivalent in volume to the passenger cabins of two jumbo jets. It is estimated that ISS will take five years to construct and will require 45 **rocket** launches. It is expected to cost more than $10 billion over its ten-year lifetime.

In November 1998, the first component of ISS was launched into space. Known as *Zarya*, it is a Russian cargo block. The second component of the station, *Unity*, was bolted onto *Zarya* on 6 December 1998. After the first occupation of *Zarya* in 2000, the crew is scheduled to increase to six by 2002.

International Ultraviolet Explorer
An artificial **satellite** used for scientific research in **ultraviolet radiation** reaching the Earth from celestial objects. It is a joint venture by the space agencies of the United States (**NASA**), Europe (**ESA**), and the United Kingdom, launched on 28 January 1978. It operates in real time and astronomers on the ground can control it directly. It has observed objects within the **Solar System**, such as **comets**, and also observed mysterious sources of **UV radiation** beyond.

interplanetary matter
Gas and **dust** thinly spread through the **Solar System**. The gas flows outwards from the Sun as the **solar wind**. Fine dust lies in the plane of the **Solar System**, scattering sunlight to cause the zodiacal light. Swarms of dust shed by **comets** enter the Earth's **atmosphere** to cause **meteor showers**.

interstellar exploration
Sending spacecraft beyond the **Solar System** to the stars beyond. In a sense, interstellar exploration has already begun, for *Pioneer* and *Voyager* space probes have travelled beyond the orbit of **Pluto** and one day, hundreds of

thousands of years hence, may drift into the planetary systems of other stars. But this is incidental to their main mission, the investigation of the outer planets.

- The main obstacle to the systematic exploration of other planetary systems is their immense distance. Even travelling at the enormous speed of 30,000 kps/18,000 mps, which is 10% of the speed of light, a probe would take 40 years to reach a nearby star 10 **light years** away. It would take 800,000 years to travel to the far side of our **Galaxy**.

- The rate of exploration would be enormously greater for **Von Neumann probes**, which would clone themselves on arrival in a planetary system, so that their numbers increased explosively as they spread across the Galaxy.

- Means of propulsion that have been suggested for starships include a continuous stream of hydrogen-bomb explosions; or a laser beam shone from Earth that could push a 'light-yacht' across space; or a 'ramjet' that would suck up **matter** as it moved, to use for fuel.

- Any conceivable manned starship would have to be crewed by astronauts prepared to spend their whole lives, and the lives of their children and grandchildren, on the voyage.

  See also: *search for extraterrestrial life, SETI.*

interstellar matter

Matter consisting of electrons, ions, atoms, molecules, and **dust** grains that fills the space between stars in our own and other **galaxies**. Over 100 different types of molecule exist in gas clouds in our **Galaxy**. Most have been detected by their radio emissions, but some have been found by the absorption lines they produce in the spectra of starlight. The most complex molecules, many of them based on carbon, are found in the dense clouds where stars are forming. They may be significant for the origin of life elsewhere in space.

- The importance of interstellar matter has been realized only within the last 50 years. It largely determines the form and development of a galaxy.

- It is most easily observable in the radio region of the **spectrum**, but was first detected optically.

- Condensations of interstellar matter are visible as **nebulae**. But over large parts of the sky it is detectable only because it dims, reddens, and polarizes the light of distant stars. It also causes a number of characteristic **absorption lines** in their spectra.

- Early radio observations by US radio engineers Karl Jansky (1905–1950) and Grote Reber (1911–   ) showed the general extent of interstellar matter. Further radio observations plotted the distribution of its most abundant constituent, neutral **hydrogen** atoms.

- Later radio observations located hydroxyl, **helium**, water, ammonia, and many other molecules, some of them quite complex.

- By mass, helium is 20–30% as abundant as hydrogen.

- All the other elements together do not amount to more than 3–5% of interstellar matter by mass.

Gas regions
Interstellar gas can be divided into neutral hydrogen (H I) and ionized hydrogen (H II) regions, though the two are in many cases intermingled. H II regions are bright emission nebulae surrounding hot stars. The terms were was introduced by Swedish-born Danish astronomer Bengt Strömgren (1908–1987), who showed that a hot star can completely ionize any gas surrounding it out to a certain distance and that the boundary between the ionized and the neutral gas is quite sharp.

Dust
Dust grains less than a micrometre in diameter, which form about 1% by mass of the interstellar matter, play an important role. They help to cool the gas and provide a surface on which molecules can form. As they are thoroughly mixed with the gas, they can be used to trace the structural details of the gas clouds. It is thought that interstellar grains are composed of graphite, silicate, and ice. Iron and silicate carbide have also been proposed. These grains can form in many ways, for example in the **photosphere** of cool **red giants**, and during **supernova** explosions.

Io
The third-largest moon of the planet **Jupiter**. It is the most volcanically active body in the **Solar System**, covered by hundreds of vents that erupt not lava but sulphur, giving Io an orange-coloured surface. The vulcanism is due to internal heating, caused by the stretching and squeezing of Io by Jupiter's gravitational field.

- In July 1995 the **Hubble Space Telescope** revealed the appearance of a 320-km/200-mi yellow spot on the surface of Io, located on the **volcano** Ra Patera. Though clearly volcanic in origin, astronomers are unclear as to the exact cause of the new spot.

- Using data gathered by the spacecraft *Galileo*, US astronomers concluded in 1996 that Io has a large metallic core.

- The *Galileo* space probe also detected a 10-megawatt beam of electrons flowing between Jupiter and Io.

- In 1997 instruments aboard the spacecraft *Galileo* measured the temperature of Io's volcanoes and detected a minimum temperature of 1,500°C/2,732°F (in comparison, Earth's hottest volcanoes reach only about 1,300°C/2,372°F).

IO: STATISTICS

Diameter	Distance from centre of planet	Period
3,630 km/2,260 mi	422,000 km/262,000 mi	1.77 days

---

ion rocket

A rocket that is driven not by the hot expanding gases produced by burning fuel, but by ions, which are electrically charged fragments of atoms. The ions are produced by breaking down atoms by heat and strong electric fields. They are then driven out of the craft's exhaust by an electric field, creating thrust. Ion rockets produce only weak thrust, but use their fuel very efficiently, and might be suitable for long interplanetary missions or **interstellar exploration**.

ionosphere

A layer of Earth's outer **atmosphere** (60–1,000 km/38–620 mi) that contains large numbers of ions – electrically charged particles – and strongly influences the way in which radio waves are propagated. It reflects some of them back to Earth, making round-the-globe radio transmission possible. The ions consist of negatively charged **electrons** knocked out of ordinary atoms by the Sun's **ultraviolet radiation**, and the positively charged incomplete atoms left behind. The British Antarctic Survey has estimated that the ionosphere is decreasing in thickness at a rate of 1 km/0.6 mi every five years, based on an analysis of data from 1960 to 1998. Global warming due to the **greenhouse effect** is the probable cause.

IRAS (Infrared Astronomy Satellite)

A joint US–UK–Dutch **satellite** launched in 1983 to survey **infrared radiation** from the sky. It studied areas of star formation, distant **galaxies** and possible embryo **planetary systems** around other stars, and discovered five new **comets** in our own **Solar System**. It operated for 10 months.

irregular galaxy

A class of **galaxy** with little structure. The two satellite galaxies of the **Milky Way**, the **Magellanic Clouds**, are both irregulars. Some galaxies previously classified as irregulars are now known to be normal galaxies distorted by tidal effects or undergoing bursts of star formation.

Isaac Newton Telescopes

A group of major telescopes operated by the British Royal Greenwich Observatory, situated at **La Palma Observatory** in the Canary Islands. They include the 4.2-m/165-in William Herschel Telescope, which in 1999 detected light reflected from a planet circling the star Tau **Boötis**, 50 **light years** from us.

isotope

Isotopes of a chemical element are forms of the element that are chemically identical, but have atoms that differ in mass. The different types of atom have the same atomic number (same number of **protons**), but contain a different number of **neutrons**. They may be stable or radioactive, naturally occurring, or synthesized. For example, the atom of ordinary **hydrogen** contains one proton and no neutrons. The isotope ^{2}H (deuterium) contains one proton and one neutron. The isotope ^{3}H (tritium) contains one proton and two neutrons. Here the superscript shows the total number of protons and neutrons in the nucleus.

- The term 'isotope' was coined by the English chemist Frederick Soddy (1877–1956), a pioneer researcher in atomic disintegration.

- Light atoms have roughly the same number of protons as neutrons. These isotopes are stable. Oxygen, for example, has eight protons in its nucleus. Its stable isotopes include ^{16}O, ^{17}O, and ^{18}O.

- Elements with high atomic mass numbers have many more neutrons than protons and are less stable. It is these isotopes that are more prone to radioactive decay. One example is ^{238}U, or uranium-238, which has 92 protons and 146 neutrons in its nucleus.

James Clerk Maxwell Telescope (JCMT)

The world's largest telescope specifically designed to observe millimetre-wavelength radiation from **nebulae**, stars, and **galaxies**. This radiation is the longest-wavelength **infrared radiation**. The telescope is located on Mauna Kea, Hawaii, and has a 15-m/50-ft diameter mirror. The JCMT is run by the United Kingdom and the Netherlands, and began operations in 1987.

jet

A stream of luminous material issuing from an active **galaxy**. Usually there are two, pointing in opposite directions. Sometimes there are brighter patches some distance from the galaxy, where the jet strikes intergalactic material and energy is released. Jets are thought to be caused by **black holes** at the centres of the active galaxies. These generate huge amounts of energy, largely in the form of **ultraviolet radiation** and **X-rays**. Streams of hot matter escape, but are focused by the black hole's **magnetic field** into beams travelling out from the black hole's magnetic poles.

Jodrell Bank

The famous site in Cheshire, UK, of the Nuffield Radio Astronomy Laboratory of the University of Manchester. Its largest instrument is the 76-m/250-ft radio dish of the **Lovell Telescope**. A 38 x 25 m/125 x 82 ft elliptical radio dish was introduced in 1964, capable of working at shorter wavelengths. These **radio telescopes** and others at Jodrell Bank are used in conjunction with smaller dishes elsewhere in Britain, up to 230 km/143 mi apart, in an array called **MERLIN**, to produce detailed maps of radio sources.

Julian date

A system of time used in astronomy in which days are numbered consecutively from noon **Greenwich Mean Time** (GMT) on 1 January 4713 BC. It is useful where astronomers wish to compare observations made over long time intervals. The Julian date (JD) at noon on 1 January 2000 was 2451545.0. The modified Julian date (MJD), defined as MJD = JD − 2400000.5, is more commonly used, because it starts at midnight GMT and the smaller numbers are more convenient.

Juno

One of the largest **asteroids**, and one of the first to be discovered. It was found by the Celestial Police, a group of astronomers in Germany who in 1800–15 set out to discover a supposed missing **planet**, thought to be orbiting the Sun between **Mars** and **Jupiter**. Although they did not discover the first asteroid, **Ceres** (discovered 1801), they discovered the second, **Pallas** (1802), and the next was Juno (1804). Its 'year' is 1,592 days. It is about 200 km/120 mi in diameter.

Jupiter

The fifth **planet** from the Sun, and the largest in the **Solar System**. It has a mass equal to 70% of all the other planets combined. It is largely composed of **hydrogen** and **helium**, liquefied by pressure in its interior, and probably with a rocky core larger than Earth. Its main feature is the Great Red Spot, a cloud of rising gases. Jupiter's strong **magnetic field** gives rise to a large surrounding magnetic 'shell', or **magnetosphere**, from which bursts of radio waves are detected.

Jupiter's atmosphere

Jupiter's **atmosphere** consists of clouds of white ammonia crystals, drawn out into belts by the planet's high speed of rotation (the fastest of any planet). Darker orange and brown clouds at lower levels may contain sulphur, as well as simple organic compounds. Farther down still, temperatures are warm, a result of heat left over from Jupiter's formation, and it is this heat that drives the turbulent weather patterns of the planet. In 1995, the *Galileo* **mission** revealed Jupiter's atmosphere to consist of 0.2% water, less than previously estimated.

JUPITER'S SATELLITES

Jupiter has 16 moons. The four largest moons, **Io**, **Europa** (which is the size of our Moon), **Ganymede**, and **Callisto**, are the **Galilean satellites**, discovered in 1610 by **Galileo**. Ganymede, which is about the size of **Mercury**, is the largest moon in the Solar System. Three small moons were discovered in 1979 by the *Voyager* space probes, as was a faint ring of dust around Jupiter's equator, 55,000 km/34,000 mi above the cloud tops.

The Great Red Spot

The Great Red Spot was first observed in 1664. It is 14,000 km/8,500 mi wide and 30,000 km/20,000 mi long, revolving anticlockwise. Its top is higher than the surrounding clouds and its colour is thought to be due to red phosphorus. The Southern Equatorial Belt in which the Great Red Spot occurs is subject to unexplained fluctuation. In 1989 it sustained a dramatic and sudden fading.

DIRECT HIT

Comet **Shoemaker-Levy** 9 crashed into Jupiter in July 1994. Impact markings were briefly visible in the atmosphere.

Jupiter's rings
Jupiter's faint rings are made up of dust from its moons, particularly the four inner moons. The discovery was made in 1998 from images taken by the *Galileo* probe.

JUPITER: STATISTICS

Average distance from Sun	Average distance from Sun (Earth = 1)	Orbital period (years)	Equatorial diameter	Mass (Earth = 1)	Tilt of equator
778 million km/ 484 million mi	5.2	11.86	142,800 km/ 88,700 mi	318	3.1°

Density (water = 1)	Escape velocity	Rotation period	Known satellites
1.34	61 kps/38 mps	9 hr 51 min	16

To remember the names of the four major (Galilean) moons of Jupiter: just remember

I eat green caterpillars
I eat good cake

The initial letters of these words are the initial letters of the moons, going outwards from the planet:

Io, Europa, Ganymede, Callisto

Keck Telescopes

The world's largest optical **telescopes**, situated on Mauna Kea, Hawaii. They are a pair of identical instruments, each weighing 300 tonnes and having a primary mirror 10 m/33 ft in diameter. The mirrors are unique in that each consists of 36 hexagonal sections, each controlled and adjusted by a computer to generate single images of the objects observed. Keck I received its first images in 1990. Keck II became operational in 1996. The telescopes are jointly owned by the California Institute of Technology and the University of California.

Kepler, Johannes (1571–1630)

German mathematician and astronomer who broke away from the ancient astronomical ideas still prevalent in his own time to formulate what are now called **Kepler's laws** of planetary motion: (1) the orbit of each **planet** is an **ellipse** with the Sun at one of the foci; (2) the radius vector of each planet sweeps out equal areas in equal times; (3) the squares of the periods of the planets are proportional to the cubes of their mean distances from the Sun. Kepler's laws are the basis of our understanding of the **Solar System**, and such scientists as Isaac **Newton** built on his ideas.

Kepler's astronomical work

- Kepler was one of the first advocates of Sun-centred cosmology, as put forward by **Copernicus**.

- Unlike Copernicus and **Galileo**, Kepler rejected the Greek and medieval belief that orbits must be circular in order to maintain the fabric of the cosmos in a state of perfection.

- Kepler discovered that the five Platonic solids (the only five regular polyhedra) can be nested inside a series of spheres whose sizes describe quite accurately (within 5%) the distances of the planets from the Sun. Kepler regarded this discovery as a divine inspiration that revealed the secret of the universe. Modern science regards the relationship as pure coincidence.

- In 1601 Kepler was bequeathed all of Tycho **Brahe's** data on planetary motion. He had already made a bet that, given Tycho's unfinished tables, he could find an accurate planetary orbit within a week. But it was five years before Kepler obtained his first planetary orbit, that of **Mars**. His analysis of Brahe's data led to the discovery of his three laws.

KEPLER'S LIFE

1571   Kepler is born in Weil der Stadt in Baden-Württemberg, Germany, and studies at Tübingen.

***c*. 1594** He becomes lecturer in astronomy at the University of Graz.

1596   He publishes a book in which he connects the sizes of the five Platonic solids with the distances of the planets from the Sun. Written in accordance with Copernican theories, it brings Kepler to the attention of all European astronomers and wins him the friendship of Brahe and Galileo.

1600   Kepler becomes assistant to Brahe in 1600.

1601   Kepler succeeds Brahe in 1601 as mathematician to the Holy Roman Emperor, Rudolph II.

1602   Kepler, who believes in astrology, wins fame with an astrological publication, *De fundamentis astrologiae*.

1604   Kepler's attention is diverted from the planets by his observation of the appearance of a new star, 'Kepler's nova'. It is the first **supernova** visible since the one discovered by Brahe in 1572.

1611   Kepler publishes a book on optics.

1612   As a Lutheran Protestant, he is expelled from Prague; he has already been expelled twice from Graz, and later is expelled from Linz, Austria, and then moves to Ulm.

1627   Kepler finally completes and publishes the Rudolphine Tables, which are based on Brahe's observations. These are the first modern astronomical tables, enabling astronomers to calculate the positions of the planets at any time in the past, present, or future. The publication also included other vital information, such as a map of the world, a catalogue of stars, and the latest aid to computation, logarithms.

1630   Kepler dies.

❛ *Ubi materia, ibi geometria.*
'Where there is matter, there is geometry. ❜

Kepler attributed remark

Kepler's laws

The three laws of planetary motion formulated in 1609 and 1619 by the German mathematician and astronomer Johannes **Kepler**. The first two laws of planetary motion were published in his book *Astronomia nova* (1609). His third law was published in his *De Harmonices mundi*. Kepler derived the laws after exhaustive analysis of numerous observations of the **planets**, especially **Mars**, made by Tycho **Brahe** without telescopic aid.

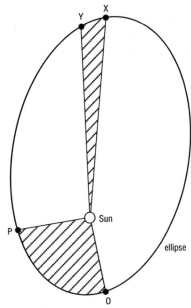

Kepler's laws *Kepler's second law states that the shaded area POS (where S is Sun) equals the shaded area XYS if the planet moves from P to O in the same time that it moves from X to Y. The law says, in effect, that a planet moves fastest when it is closest to the Sun.*

- Kepler also suggested that the Sun itself rotates, a theory that was confirmed by **Galileo's** observations of **sunspots**.

- Kepler suggested that the Sun's rotation set up some sort of 'magnetic' interaction between the Sun and the **planets**, driving them in orbit. This idea, although incorrect, was an important precursor of Isaac **Newton's** theory of **gravity**.

- Newton later showed that Kepler's Laws were a consequence of his theory of universal gravitation.

KEPLER'S LAWS

- The first law stated that planets travel in elliptical orbits rather than in circles or combinations of circles. The Sun occupies one of the two foci of each **ellipse**.
- The second law established the Sun as the main force governing the orbits of the planets. It stated that the line joining the Sun and a planet traverses equal areas of space in equal periods of time, so that the planets move more quickly when they are nearer the Sun.
- The third law described in precise mathematical language the link between the distances of the planets from the Sun and their velocities – specifically, that the squares of the periods of the planets are proportional to the cubes of their mean distances from the Sun.

Kirkwood gaps
Regions of the **asteroid** belt, between **Mars** and **Jupiter**, where there are relatively few asteroids.

- The orbital periods of particles in the gaps correspond to simple fractions, especially $\frac{1}{3}$, $\frac{2}{5}$, $\frac{3}{7}$, and $\frac{1}{2}$, of the orbital period of Jupiter, indicating that they are caused by the gravitational influence of the larger **planet**.

- The gaps are named after Daniel Kirkwood (1814–1895), the 19th-century US astronomer who first drew attention to them.

Kitt Peak National Observatory
An **observatory** in the Quinlan Mountains near Tucson, Arizona, operated by AURA (Association of Universities for Research into Astronomy). Its main **telescopes** are the 4-m/158-in Mayall reflector, opened in 1973, and the McMath Solar Telescope, opened in 1962, the world's largest of its type. Among numerous other telescopes on the site is a 2.3-m/90-in reflector owned by the Steward Observatory of the University of Arizona.

Kourou
Site of the Guiana Space Centre of the **European Space Agency**. Kourou is the second-largest town of French Guiana, north-west of Cayenne, population (1996) 20,000 (20% of the total population of French Guiana). Situated near the equator, it is an ideal site for launches of **satellites** into **geostationary orbit**.

Kuiper Belt
A ring of small, icy bodies orbiting the Sun beyond the outermost **planet**, **Pluto**. The Kuiper Belt, named after the US astronomer Gerard Kuiper (1905–1973), who proposed its existence in 1951, is thought to be the source of **comets** that orbit the Sun with periods of less than 200 years. The first member of the Kuiper Belt was detected in 1992. In 1995 the first comet-sized objects were discovered; previously the only objects found had diameters of at least 100 km/63 mi (comets generally have diameters of less than 10 km/6.3 mi).

La Palma Observatory
Major astronomical **observatory**, also called Observatorio del Roque de los Muchachos, located on La Palma, most westerly of the Canary Islands, Spain. It is the home of the **Isaac Newton telescopes**, including the 4.2-m/165-in William Herschel Telescope. These instruments are operated by the British **Royal Greenwich Observatory**. The La Palma Observatory began operating in 1987.

Lagrange point
Five locations in space where the gravitational forces of two bodies neutralize each other, so that a third, less massive body located at any one of these points will tend to stay near it. Three of the points, L1–L3, lie on a line joining the two large bodies. The other two points, L4 and L5, which are the most stable, lie on either side of this line. Their existence was predicted in 1772 by the French mathematician Joseph Louis Lagrange (1736–1813). The **Trojan asteroids** lie at Lagrangian points L4 and L5 in **Jupiter's** orbit around the Sun. Clouds of **dust** and debris may lie at the Lagrangian points of the Moon's orbit around the Earth.

latitude
A measure of position north or south of the equator. Lines of latitude are imagined drawn parallel to the equator, with 0° at the equator and 90° at the north and south poles. The 0° line of latitude is defined by the equator, and can be found by astronomical observation. It was determined as early as AD 150 by the Egyptian astronomer **Ptolemy** in his world atlas.
   See also: *longitude.*

Leavitt, Henrietta Swan (1868–1921)
US astronomer who in 1912 discovered a key law that helps to establish the distance scale within our **Galaxy** and beyond it to the nearby **galaxies**.

- Leavitt discovered the period–luminosity law, which links the brightness of a Cepheid variable star to its **period** of variation.

- By comparing a Cepheid's apparent **magnitude** with its absolute magnitude, found by applying the law to the measured period of variation, the distance of the star from the Earth could be deduced.

- The period–luminosity law therefore allows astronomers to use **Cepheid variables** as 'standard candles' for measuring distances in space.

- Leavitt's period–luminosity curve for the Cepheids made it possible to develop the **Hertzsprung–Russell diagram.**

- Leavitt also discovered a total of 2,400 new **variable** stars and four **novae.**

Leo

A **zodiacal constellation** in the northern hemisphere, represented as a lion. The Sun passes through Leo from mid-August to mid-September. Its brightest star is first-**magnitude Regulus** at the base of a pattern of stars called the Sickle. In astrology, the dates for Leo are between about 23 July and 22 August.
   See also: *precession.*

Leverrier, Urbain Jean Joseph (1811–77)

French astronomer who correctly predicted the existence and position of the planet **Neptune** from its influence on the orbit of the planet **Uranus**, which was discovered in 1846. The possibility that another **planet** might exist beyond Uranus, influencing its orbit, had already been suggested. Leverrier calculated the **orbit** and probable position of the hypothetical planet, and wrote to a number of **observatories**, asking them to test his prediction of its position. Johann Galle (1812– 1910) at the Berlin Observatory found it immediately, within 1° of Leverrier's coordinates.

Libra

A faint **zodiacal constellation** on the celestial equator (see **celestial sphere**) adjoining **Scorpius**, and represented as the scales of justice. The Sun passes through Libra during November. The constellation was once considered to be a part of Scorpius, seen as the scorpion's claws. In astrology, the dates for Libra are between about 23 September and 23 October.
   See also: *precession.*

libration

A slight apparent 'wobble' in the rotation of the Moon. It is due to the Moon's variable orbital speed and the tilt of its axis.

- Generally, the Moon rotates on its axis in the same time as it takes to complete one orbit, causing it to keep one face turned permanently towards the Earth. Its speed in orbit varies, however, because its **orbit** is not circular but elliptical (see **Kepler's laws**). So at times the Moon's axial rotation appears to get either slightly ahead of or slightly behind its orbital motion, so that part of the far side of the Moon is visible around the east and west edges. This is known as libration in **longitude.**

- Libration in **latitude** occurs because the Moon's axis is slightly tilted with respect to its orbital plane, so we can see over the north and south poles.

- In combination, these effects mean that a total of 59% of the Moon's surface is visible from the Earth.

life beyond the Earth

The study of life forms that may exist elsewhere in the universe is called **exobiology**. The techniques at its disposal range from space-probe experiments designed to detect organic molecules to the monitoring of radio waves from other star systems. Human excursions into space have shown that conditions on Earth at the time life began are not unique to this planet. Similar conditions prevail elsewhere in the **Solar System**. But complex life-forms cannot have developed elsewhere in the Solar System, though they may exist on other planets within our **Galaxy**.

See also: *life from space, origin of life, SETI.*

 Even if life did exist elsewhere, we might not recognize it. Absence of evidence won't be evidence of absence.

Martin Rees, British Astronomer Royal, *The Independent*, 18 January 1997

POINTERS TO LIFE?

- Traces of carbon compounds have been found in **interstellar matter** – suggesting that life could form in space, or else that planetary surfaces could be 'seeded' from space with the building blocks of life.

- Space probes located water and ice in the polar regions of the **Moon** in 1998. Scientists don't expect to find life on the Moon, but this is a hopeful sign that water may be common on more hospitable worlds around other stars.

- Beneath the ice that covers **Europa**, one of the four large **moons** of **Jupiter**, there may be relatively warm oceans of water. This is another place in which to look for signs of life.

- There is liquid water at the poles of **Mars** for at least part of the Martian 'year'. Micro-organisms could have developed here.

life from space

The idea that terrestrial life did not begin on Earth but came here from space has been proposed by the British astrophysicist and cosmologist Sir Fred **Hoyle**, with his colleague the Indian astrophysicist Chandra Wickramasinghe. They suggest that complex molecules built up in interstellar gas clouds, and are scattered on the surfaces of **planets**, moons, and other celestial bodies by **comets**. There they can evolve into complex forms. Hoyle and Wickramasinghe further suggest that epidemics may be caused by visiting comets, or by the Earth passing through debris left by them. Other scientists have looked for traces of life from space in **meteorites**.

VISITORS FROM MARS?

Repeated claims have been made that traces of biological processes can be found in meteorites. There was great public excitement over the 1996 report by a team of **NASA** scientists that they had found complex organic molecules in meteorites found in Antarctica. These meteorites almost certainly come from **Mars**, thrown off by a past impact with an **asteroid** or comet. But the claims that they show traces of life have been challenged by other scientists and are still regarded as unproved.

---

light

Light consists of electromagnetic waves in the visible range of wavelength, from about 400 nanometres in the extreme violet to about 770 nanometres in the extreme red. But light shows not only wave but also particle properties. The fundamental particle, or quantum, of light is called the photon. The speed of light (and of all electromagnetic radiation) in a vacuum is precisely 299,792.458 kps (approximately 186,282.40 mps). It is symbolized by c.

See also: *diffraction, refraction.*

light year

The distance travelled by **light** in a vacuum in one year. It is approximately 9.46 trillion (million million) km/5.88 trillion miles. It is commonly used as the unit of distance to stars and **galaxies**, though professional astronomers prefer the **parsec**.

limb

The edge of the visible disc of the Sun, Moon, a **planet**, and so on.

Local Group

A cluster of about 30 **galaxies** that includes our own, the **Milky Way**. Like other groups of galaxies, the Local Group is held together by the gravitational attraction among its members, and does not expand with the expanding universe.

- The Local Group contains at least 21 members within a region 3 million light years across. The two largest galaxies are the Milky Way and the **Andromeda galaxy**, two large Sb spirals. Most of the others are small and faint. They include one smaller Sc spiral, 14 ellipticals, of which ten are dwarfs, and four irregulars, of which two are the **Magellanic Clouds**.

- Astronomers keep finding small, faint galaxies that are nearby but partly hidden by our own Milky Way. One recent discovery is named MB1, and is a **spiral galaxy** 17,000 **light years** across. Another, MB2, is an irregular-shaped **dwarf galaxy** about 4,000 light years across.

- In 1996 US astronomers discovered a further new galaxy 17 million light years away.

- It has been suggested that our Local Group is only a sub-section of a **Local Supercluster**.

Local Supercluster

A hypothetical grouping of clusters of **galaxies**, including our own **Local Group**. If it exists, this supercluster may be centred in or near the large cluster in **Virgo** and is about 100 million light years in diameter and 25 million light years thick.

longitude

A measure of position east or west of an arbitrarily defined **meridian**. Imaginary lines of longitude are drawn running through both poles, with 0° (the prime meridian) passing through Greenwich, London, UK. The prime meridian is a matter of convention rather than physics.

Lovell Telescope

Giant **radio telescope** at the Nuffield Radio Astronomy Laboratory at **Jodrell Bank** in Cheshire, UK. It was completed in 1957 and immediately became famous when it proved well suited to tracking *Sputnik*, the first artificial **satellite**. It was substantially modified in 1970. The dish is 76 m/250 ft across.

luminosity

The true brightness of a celestial object, in terms of the amount of radiant energy that it emits each second. It can be expressed in terms of absolute **magnitude**.

Luna missions
A series of 24 crewless Soviet space probes launched towards the Moon from 1959 to 1976.

- *Luna 2* was the first object from Earth to hit the Moon, in September 1959.

- *Luna 3* passed behind the Moon in October 1959 and took the first pictures of the Moon's farside.

- *Luna 9* was the first spacecraft to make a soft landing on the Moon, in January 1966.

- *Luna 17* (November 1970) and *Luna 21* (January 1973) landed *Lunokhod* roving vehicles.

- *Luna 16* (September 1970), *Luna 20* (February 1972), and *Luna 24* (August 1976) returned soil samples to Earth; no spacecraft has soft-landed on the Moon since.

lunar eclipse
A temporary darkening of the Moon that occurs when it passes into the shadow of the Earth, becoming dim until emerging from the shadow. Lunar eclipses may be partial or total, and they can happen only at full Moon. Total lunar eclipses last for up to 100 minutes; the maximum number each year is three.

See also: *eclipse, solar eclipse.*

Lyra
A small but prominent **constellation** of the northern hemisphere, represented as the lyre of the musician Orpheus of Greek mythology.

- Lyra's brightest star is **Vega**.

- Epsilon Lyrae is a system of four gravitationally linked stars.

- Beta Lyrae is an **eclipsing binary**.

- The Ring nebula, M57, is a planetary **nebula**.

Magellan
A **NASA** space probe to **Venus**, launched in May 1989. It went into orbit around Venus in August 1990 to make a detailed map of the **planet** by radar. It revealed **volcanoes**, meteoric **craters** and fold mountains on the planet's surface. Magellan mapped 98% of Venus. In October 1994 Magellan was commanded to self-destruct by entering the **atmosphere** around Venus, where it burned up.

Magellanic Clouds
The two **galaxies** nearest to our own. They are irregularly shaped, and look like detached parts of the **Milky Way**, lying in the southern **constellations** Dorado, Tucana, and Mensa.

- The Large Magellanic Cloud (LMC) spreads over the constellations of Dorado and Mensa.

- The LMC is 169,000 **light years** from Earth, and about a third the diameter of our **Galaxy**.

- The Small Magellanic Cloud (SMC) is in Tucana.

- The SMC is 180,000 light years away, and is about a fifth the diameter of our Galaxy.

- Being the nearest galaxies to ours (the distance to the **Andromeda galaxy** is 2,200,000 light years), the Clouds are especially useful for studying stellar populations and objects such as **supergiant** stars.

- It was for **Cepheid variables** in the SMC that the **period**–luminosity relationship was first established by Henrietta **Leavitt** in 1912.

- The Magellanic Clouds are named after the Portuguese navigator Ferdinand Magellan (c. 1480–1521), who was the first European to describe them.

magnetic field
The region around a magnet, or around a conductor carrying an electric current, in which a force acts on a moving electric charge or on a magnet (which at the atomic level consists of moving charges). The Earth has a magnetic field, which is generated by currents in its liquid iron core. At the

surface, the field orients the needles of magnetic compasses and magnetizes rocks as they form, so that the direction and strength of the magnetic field of ancient times can be discovered by measuring the magnetism of rocks formed at those times. In near space the Earth's field traps charged particles from the **solar wind** in the **radiation belts**. The magnetic fields of **Mercury** and **Mars** are much weaker than the Earth's; those of **Jupiter** and **Saturn** are far stronger. The **Sun's** magnetic field controls the formation of sun**spots** and the motion of **prominences** and **flares**. Other stars have magnetic fields. Intense magnetic fields are associated with **pulsars** and with the turbulent cores of **active galaxies**.

See also: *aurora, magnetic storm, magnetosphere.*

magnetic storm

A sudden disturbance affecting the Earth's **magnetic field**, causing anomalies in radio transmissions and magnetic compasses. It is caused by solar activity.

See also: *aurora, radiation belts, sunspot.*

magnetosphere

The volume of space surrounding a **planet** that is controlled by the planet's **magnetic field**. The Earth's magnetosphere extends 64,000 km/40,000 mi towards the Sun, but many times this distance on the side away from the Sun. The extension away from the Sun is called the magnetotail. The outer edge of the magnetosphere is the magnetopause. Beyond this is a turbulent region, the magnetosheath, where the **solar wind** is deflected around the magnetosphere. Inside the magnetosphere, atomic particles follow the Earth's lines of magnetic force. The magnetosphere contains the Van Allen **radiation belts**. Other planets have magnetospheres, notably **Jupiter**.

magnitude

A measure of the brightness of a star or other celestial object. The larger the number denoting the magnitude, the fainter the object.

- Zero or first magnitude indicates some of the brightest stars.
- Still brighter are those of negative magnitude, such as **Sirius**, whose magnitude is −1.46.
- *Apparent magnitude* is the brightness of an object as seen from Earth.
- *Absolute magnitude* is the brightness at a standard distance of 10 **parsecs** (32.6 **light years**).
- Each magnitude step is equal to a brightness difference of 2.512 times. Thus a star of magnitude 1 is $(2.512)^5$ or 100 times brighter than a sixth-magnitude star, which is just visible to the naked eye.

- The apparent magnitude of the Sun is –26.8.

- The absolute magnitude of the Sun is 4.8 – it would look like a faint star at the standard distance of 32.6 light years.

main sequence

The part of the **Hertzsprung–Russell diagram** that contains most of the stars, including the Sun. It runs diagonally from the top left of the diagram to the lower right. The most massive (and hence brightest) stars are at the top left, with the least massive (coolest) stars at the bottom right. The main sequence consists of stars that are shining by the fusion of **hydrogen** into **helium**. When its hydrogen fuel runs low, a star evolves upwards, away from the main sequence, into the region occupied by **giant** stars and **supergiants**.

Manned Manoeuvrability Unit (MMU)

A **NASA** device that enables an astronaut to move about outside a spacecraft without being tethered. It is in fact a personal jet-pack, though it resembles a flying armchair in appearance.

- An MMU was first used by astronaut Bruce McCandless on the tenth **Space Shuttle** mission in February 1984.

- On several later missions MMUs were used in retrieving damaged **satellites** and pulling them into the Shuttle's cargo bay.

- MMUs have not been used since the *Challenger* Shuttle disaster of January 1986 – NASA has become very cautious – but they, or something similar, will be required during the building of the **International Space Station**.

mare

Any of the numerous dark lowland plains on the Moon. The name is the Latin word for 'sea', because these areas were once wrongly thought to be water. Its plural is 'maria'.

Mariner spacecraft

A series of US space probes that explored the planets **Mercury**, **Venus** and **Mars** from 1962 to 1975.

See also: *Venera missions, Voyager missions.*

Mariner missions

Mission number	Year	Notes
1	1962	To Venus; failed launch.
2	1962	First fly-by of Venus, at 34,000 km/21,000 mi; confirmed the existence of the solar wind, and measured Venusian temperature.

Mariner missions (*continued*)

Mission number	Year	Notes
3	1964	Did not achieve its intended trajectory to Mars.
4	1965	Passed Mars at a distance of 9,800 km/6,100 mi, and took photographs, revealing a dry, cratered surface.
5	1967	Passed Venus at 4,000 km/2,500 mi, and measured Venusian temperature, atmosphere, mass, and diameter.
6, 7	1969	Photographed the Martian equator and southern hemisphere respectively, and also measured temperature, atmospheric pressure and composition, and diameter.
8	1971	To Mars; failed launch.
9	1971	Mapped the entire Martian surface, and photographed the two moons. Its photographs revealed the changing of the polar caps, and the extent of vulcanism, canyons, and features suggesting that there might once have been water on Mars.
10	1974–75	Took close-up photographs of Mercury and Venus, and measured temperature, radiation, and magnetic fields.
11, 12	–	*Mariner 11* and *12* were renamed *Voyager 1* and *2*.

Mars

The fourth **planet** from the Sun. It is much smaller than **Venus** or Earth, with a mass only one-ninth that of Earth. Mars is slightly pear-shaped, with a low, level northern hemisphere, which is comparatively uncratered and geologically 'young', and a heavily cratered 'ancient' southern hemisphere. Mars has two small **satellites**, **Phobos** and **Deimos**.

Martian Surface
- The landscape is a dusty, red, eroded lava plain.
- Mars has white polar caps (water ice and frozen carbon dioxide) that advance and retreat with the seasons
- There are four enormous **volcanoes** near the equator, of which the largest is Olympus Mons, 24 km/15 mi high, with a base 600 km/375 mi across, and a **crater** 65 km/40 mi wide.
- To the east of the four volcanoes lies a high plateau cut by a system of valleys, Valles Marineris, some 4,000 km/2,500 mi long, up to 200 km/120 mi wide and 6 km/4 mi deep; these features are apparently caused by faulting and wind erosion.
- Recorded temperatures vary from –100°C/–148°F to 0°C/32°F.
- Studies in 1985 showed that enough water might exist to sustain prolonged missions by space crews.

- Photographs from the *Mars Pathfinder* indicated that the planet is rusting. **NASA** announced this in July 1997 and said that a supercorrosive force was eroding rocks on the surface, owing to iron oxide in the soil.

- Images from the **Hubble Space Telescope** showed that dust storms had covered areas of the planet that had been dark features in the early century, including one section as large as California.

MARTIAN ATMOSPHERE

- The surface atmospheric pressure is less than 1% of the Earth's atmospheric pressure at sea level.
- The atmosphere consists of 95% carbon dioxide, 3% nitrogen, 1.5% argon, 0.15% oxygen.
- Red atmospheric dust from the surface, whipped up by winds of up to 450 kph/280 mph, causes the sky to be light pink.
- In May 1997 American scientists announced that Mars is becoming increasingly colder and cloudier.

Missions to Mars

- The first human-made object to orbit another planet was *Mariner 9* (see **Mariner** missions).

- **Viking** *1* and *Viking 2*, each of which consisted of an orbiter and a lander, provided vast amounts of information and photographs from the surface.

- In January 1997 NASA launched the *Mars Pathfinder*, which made a successful landing on Mars in July 1997 on a flood plain called Ares Vallis.

- After initial technical problems, *Pathfinder's* 0.3-m/1-ft rover, *Sojourner*, began to explore the Martian landscape and to transmit data back to Earth.

- The *Global Surveyor*, which entered Martian orbit in September 1997, revealed that Mars's **magnetic field** is a mere 800th that of the Earth.

- NASA's Mars *Climate Orbiter* to monitor weather on Mars was launched from Cape Canaveral, Florida, in December 1998. It was expected to reach its destination in September 1999. However, a measurement error caused the probe to fly too close to Mars and break up.

- NASA launched its Mars *Polar Lander* in January 1999. The spacecraft was intended to send a *Polar Lander* down by parachute to the south pole, the first probe in that area. It carried a microphone and would have been the first planetary probe to send back sounds from another world. However, the probe was lost during its final approach to Mars.

- **The European Space Agency's** (ESA) *Mars Express* was given the go-ahead in November 1998. It is scheduled to be launched in 2003. It will use radar to map possible underground water sources and will also carry a lander, *Beagle 2*, for soil collection.

MARS: STATISTICS

Average distance from Sun	Average distance from Sun (Earth = 1)	Orbital period (years)	Equatorial diameter	Mass (Earth = 1)	Tilt of equator
227.9 million km/ 141.6 million mi	1.52	1.88	6,790 km/ 4,219 mi	0.1075	23° 59'

Density (water = 1)	Escape velocity	Rotation period	Known satellites
3.95	5.0 kps/3.1 mps	24h 37m 22.6s	2

Mars Pathfinder
A US spacecraft that landed in the Ares Vallis region of **Mars** on 4 July 1997. It carried a small six-wheeled roving vehicle called *Sojourner* which examined rock and soil samples around the landing site. *Mars Pathfinder* was the first to use air-bags instead of retro-rockets to cushion the landing.

mascon
One of a number of regions on the surface of the Moon where **gravity** is stronger than average. The effect is due to localized areas of denser rock strata. The name is a contraction of 'mass concentration'.

mass transfer in binary star systems
Complex exchanges of matter can happen between the members of a **binary star** system towards the end of their lives. When one of the stars exhausts the **hydrogen** fuel in its core, it swells to become a **giant** star. Matter in its outer layers, largely consisting of unburned hydrogen, is captured by the **gravity** of its partner, and is sucked onto the surface. This can cause a flare-up of the second star, turning it into an **eruptive variable**, a **nova**. It may itself now age rapidly and swell, and in turn lose matter to its partner, temporarily rejuvenating it. One type of **supernova** consists of stars in binary systems that have received sudden transfusions of matter in this way.

mass–luminosity relationship

A mathematical relationship between the mass of a star and its **luminosity**, or true brightness. The more mass a star has, the faster it burns its **hydrogen** fuel, and the brighter it will shine.

Mauna Kea

A dormant **volcano** in Hawaii, USA, site of a major astronomical **observatory** at 4,200 m/13,784 ft above sea level. Because of its elevation high above clouds, atmospheric moisture, and artificial lighting, Mauna Kea is ideal for **infrared astronomy**. The first **telescope** on the site was installed in 1970. Today telescopes include the:

- 2.24-m/88-in University of Hawaii reflector (1970)

- 3.8-m/150-in United Kingdom Infrared Telescope (UKIRT), also used for optical observations (1979)

- 3-m/120-in NASA Infrared Telescope Facility (IRTF) (1979)

- 3.6-m/142-in Canada–France–Hawaii Telescope (CFHT), designed for optical and infrared work (1979)

- 15-m/50-ft UK/Netherlands **James Clerk Maxwell Telescope** (JCMT), the world's largest telescope specifically designed to observe millimetre-wave radiation from **nebulae**, stars, and **galaxies** – the JCMT is operated via satellite links by astronomers in Europe (1987)

- world's largest optical telescopes, the twin **Keck Telescopes**.

mean Sun

An imaginary Sun that moves around the sky during the year at a constant rate equal to the average rate of the real Sun. The real Sun speeds up and slows down as viewed by us because the Earth moves at a variable rate in its orbit, so that the time shown by a sundial is fast or slow of the time shown by an accurate clock. Local·mean **solar time** is the time in relation to the mean Sun.

See also: *equation of time, Greenwich mean time.*

Mercury

The closest **planet** to the Sun. It has only thin traces of an **atmosphere**, which offers no protection against the heat of the Sun on the daytime side or the chill of space on the night side. On its sunward side the surface temperature reaches over 400°C/752°F, but on the 'night' side it falls to −170°C/−274°F.

Mercury facts

- Mercury's atmosphere contains minute traces of argon and **helium**.
- The surface is composed of silicate rock, often in the form of lava flows.
- In 1974 the US space probe *Mariner 10* showed that Mercury's surface is cratered by **meteoroid** impacts.
- In 1999 **NASA** approved a $286-million mission to send the spacecraft *Messenger* into orbit around Mercury in 2008 (after being launched in 2004), to photograph the planet's surface, analyse its atmospheric composition, and map its **magnetic field.**
- Mercury's largest feature is the Caloris Basin, 1,400 km/870 mi wide.
- There are also cliffs hundreds of kilometres long and up to 4 km/2.5 mi high, thought to have been formed by the cooling of the planet billions of years ago.
- Inside the planet is an iron core three-quarters of the planet's diameter, which produces a magnetic field 1% of the strength of Earth's.

MERCURY: STATISTICS

Average distance from Sun	Average distance from Sun (Earth = 1)	Orbital period	Equatorial diameter	Mass (Earth = 1)	Tilt of equator
58 million km/ 36 million mi	0.387	88 days	4,880 km/ 3,030 mi	0.056	2°

Density (water = 1)	Escape velocity	Rotation period	Known satellites		
5.4	4.2 kps/2.6 mps	59 days	0		

Mercury missions

A US project to put a human being in space, in the one-seat *Mercury* spacecraft, conducted from 1961 to 1963. The first two *Mercury* flights, launched by Redstone **rockets**, were short flights to the edge of space and back. The orbital flights, beginning with the third in the series (made by John **Glenn**), were launched by Atlas rockets.

meridian

1   Half of a great circle drawn on the Earth's surface, passing through both poles and thus through all places with the same **longitude**. Terrestrial longitudes are usually measured from the Greenwich **Meridian**.

2   A great circle on the **celestial sphere** passing through the celestial pole and the **zenith** (the point immediately overhead).

MERLIN

An array of radio **telescopes** located in Britain and including instruments at the **Jodrell Bank** observatory, Cheshire. The radio telescopes make coordinated observations of radio sources in the sky, gaining much more detailed information about the structure of what they are observing than any single instrument could. The name is an acronym of 'Multi-Element Radio-Linked Interferometer Network'.

mesosphere

A layer of the Earth's **atmosphere** lying between the **stratosphere** and **exospher**, at between 50 km/31 mi and 80 km/50 mi high.

- In the mesosphere temperature decreases with altitude, from 0°C/32°F to below −100°C/−148°F.

- Ultraviolet light knocks **electrons** out of atoms to create a plasma, a gas of atom fragments: negatively charged electrons and positively charged ions.

- The resulting **ionosphere** acts as a reflector of radio waves, enabling radio transmissions to 'hop' between widely separated points on the Earth's surface.

Messier catalogue

A catalogue of 103 **galaxies**, **nebulae**, and star clusters (the Messier objects) published in 1784 by the French astronomer Charles Messier (1730–1817). Catalogue entries are denoted by the prefix 'M'. Well known examples include M31 (the **Andromeda galaxy**), M42 (the **Orion Nebula**), and M45 (the Pleiades **open cluster**). The list was later extended to 109.

Some of the most spectacular objects in the Messier catalgoue

Messier number	Name	Constellation	Notes
1	Crab Nebula	Taurus	Supernova remnant
8	Lagoon Nebula	Sagittarius	Diffuse nebula
17	Omega Nebula	Sagittarius	Diffuse nebula
20	Trifid Nebula	Sagittarius	Diffuse nebula
27	Dumbbell Nebula	Vulpecula	Planetary nebula
31	Andromeda Galaxy	Andromeda	Spiral galaxy

Some of the most spectacular objects in the Messier catalgoue (*continued*)

Messier number	Name	Constellation	Notes
42	Orion Nebula	Orion	Diffuse nebula
44	Praesepe	Cancer	Open cluster
45	Pleiades	Taurus	Open cluster
51	Whirlpool Galaxy	Canes Venatici	Spiral galaxy
57	Ring Nebula	Lyra	Planetary nebula

THE COMET FERRET

▨ Charles Messier watched for the predicted return of Halley's **comet**, and was one of the first people to spot it. The experience inspired him with the desire to go on discovering new comets for the rest of his life. Louis XV nicknamed him the 'Comet Ferret'.

▨ Messier's search was continually hampered by fuzzy objects, which, however, could not be comets since they did not move against the background of fixed stars. During the period 1760–84, therefore, he compiled a list of these nebulae and star clusters, so that he and other astronomers would not confuse them with possible new comets.

▨ Joseph Lalande (1732–1807) named a **constellation** after Messier, but the name has fallen into disuse.

meteor

A point of light darting across the sky, sometimes leaving a briefly glowing trail, popularly known as a shooting or falling star. It is caused by a particle of **dust**, a **meteoroid**, entering the **atmosphere** at speeds up to 70 kps/45 mps and burning up by friction, at a height of around 100 km/60 mi. On any clear night, several sporadic meteors can be seen each hour. A brilliant meteor is termed a fireball.

meteor *Long exposure photograph of circumpolar stars with a meteor trail diagonally across the photograph.*

Meteor Crater

Impact **crater** near Winslow, Arizona, caused by the impact of a

50-m/165-ft iron **meteoroid** some 25,000 years ago. It is 1.2 km/0.7 mi in diameter, 200 m/660 ft deep and the walls are raised 50–60 m/ 165–198 ft above the surrounding desert. It is also called Coon Butte, or Barringer Crater. The last name comes from US mining engineer, Daniel Barringer, who proposed in 1902 that the feature was an **impact crater** rather than volcanic, an idea confirmed in the 1960s by US geologist and astronomer Eugene Shoemaker.

Meteor Crater *The Arizona Meteor Crater.*

meteor shower

A number of **meteors**, ranging from dozens to tens of thousands, seen over a period of hours or even several nights. Several times each year the Earth encounters swarms of **dust** shed by **comets**, which give rise to meteor showers. Each appears to radiate from one particular point in the sky, after which the shower is named; for example, the meteors in the Perseid shower in August appear to move away from the **constellation Perseus**. The Leonids radiate from the constellation **Leo** and are caused by dust from Comet Tempel–Tuttle, which orbits the Sun every 33 years. The Leonid shower reaches its peak when the comet is closest to the Sun.

Major night-time meteor showers*

Duration	Name	Constellation in which radiation lies
1–6 Jan	Quadrantids	Boötes
19–24 April	April Lyrids	Lyra
1–8 May	Eta Aquarids	Aquarius (near star Eta Aquarii)
15 July–15 August	Delta Aquarids	Aquarius (near star Delta Aquarii)
25 July–18 August	Perseids	Perseus
16–26 Oct	Orionids	Orion
20 Oct–30 Nov	Taurids	Taurus
15–19 Nov	Leonids	Leo
7–15 Dec	Geminids	Gemini

*A shower's peak rate occurs around the middle of the period shown.

meteorite

A piece of rock or metal from space that reaches the surface of the Earth, Moon, **Mars** or other celestial body. It is the remnant left by a **meteor**. Most meteorites are thought to be fragments from **asteroids**, although some may be pieces from the heads of **comets**. Most are stony, although some are made of iron and a few have a mixed rock–iron composition.

meteorite *Piece of meteorite which fell in the English Midlands in the late 1960s.*

- Stony meteorites can be divided into two kinds: **chondrites** and **achondrites**.

- Chondrites contain **chondrules**, small spheres of the silicate minerals olivine and orthopyroxene, and comprise 85% of meteorites. Achondrites do not contain chondrules.

- Meteorites provide evidence for the nature of the **Solar System** and may be similar to the Earth's core and mantle, neither of which can be observed directly.

- Thousands of meteorites hit the Earth each year, but most fall in the sea or in remote areas and are never recovered.

- The largest known meteorite is composed of iron, weighing 60 tonnes, which lies where it fell in prehistoric times at Grootfontein, Namibia, in southern Africa.

- Meteorites are slowed down by the Earth's **atmosphere**, but if they are moving fast enough they can form a **crater** on impact.

Heaviest meteorites

(n/a = not available)

Name and location	Weight (tonnes)	Year found	Dimensions	Composition
Hoba West, Grootfontein, Namibia	60	1920	2.7 x 2.7 x 1.1 m 9 x 9 x 3.5 ft	nickel-rich iron
Ahnighito, Greenland	30	n/a	n/a	n/a
Bacuberito, Mexico	27	1863	3.7 m/12 ft long	iron
Mbosi, Tanzania	26	1930	13.5 x 4 x 4 ft 4.1 x 1.2 x 1.2 ft	iron
Agpalik, Greenland	20	n/a	n/a	n/a
Armanty, Mongolia	20	1935 (known in 1917)	n/a	iron

> **NEAR MISS**
>
> An explosion in Northern Ireland in December 1997 was blamed on terrorists, but was later discovered to be caused by a meteorite. It left a 1.2-m/4-ft wide crater.

meteoroid
A grain of dust or larger piece of rock, moving in interplanetary space.

- There is no official distinction between meteoroids and **asteroids**, except that the term 'asteroid' is generally reserved for objects more than 1.6 km/1 mi in diameter, whereas meteoroids can range from the size of a grain of dust upwards.

- Some meteoroids result from the fragmentation of asteroids after collisions, others are shed by **comets**.

- Some meteoroids strike the Earth's **atmosphere**, and their fiery trails are called **meteors**.

- If some of the meteoroid reaches the ground, the remnant is named a **meteorite**.

- The Earth sweeps up an estimated 16,000 tonnes of meteoric material every year.

micrometeoroids
Tiny **meteoroids**, so easily slowed down by friction if they enter the Earth's **atmosphere** that they do not burn up to produce a luminous **meteor**. They merely float down through the atmosphere, and many can be found at high altitudes.

Milky Way
A faint band of light crossing the night sky, consisting of stars in the plane of our **Galaxy**. The name 'Milky Way' is often used for the Galaxy itself.

- The densest parts of the Milky Way, towards the Galaxy's centre, lie in the constellation **Sagittarius**.

- In places, the Milky Way is interrupted by lanes of dark dust that obscure light from the stars beyond, such as the Coalsack Nebula in **Crux** (the Southern Cross).

- It is because of these dark dust-clouds that the Milky Way is irregular in width and appears to be divided into two between **Centaurus** and **Cygnus**.

- The Milky Way passes through the **constellations** of **Cassiopeia**, **Perseus**, **Auriga**, **Orion**, **Canis Major**, Puppis, Vela, Carina, Crux, Centaurus, Norma, **Scorpius**, Sagittarius, Scutum, **Aquila**, and Cygnus.

Mir

A Soviet **space station**, the core of which was launched on 20 February 1986. It was permanently occupied until 1999. Its name is a Russian word meaning 'peace' or 'world'.

- In June 1995 the US **space shuttle** *Atlantis* docked with *Mir*, exchanging crew members.

- A small wheat crop was harvested aboard *Mir* on 6 December 1996. It was the first successful cultivation of a plant from seed in space.

- *Mir* weighs almost 21 tonnes, is approximately 13.5 m/44 ft long, and has a maximum diameter of 4.15 m/13.6 ft.

Mira or Omicron Ceti

The brightest long-period pulsating **variable** star, located in the **constellation Cetus.** Mira was the first star discovered to vary periodically in brightness.

- In 1596 a Dutch astronomer, David Fabricus, noticed Mira as a third-**magnitude** star. Because it did not appear on any of the star charts available at the time, he mistook it for a **nova**.

- The German astronomer Johann Bayer (1572–1625) included it on his star atlas in 1603 and designated it Omicron Ceti.

- The star vanished from view again, only to reappear within a year. It was named 'Stella Mira', 'the wonderful star', by Johannes Hevelius (1611–1687), who observed it over the period 1659–82.

- Mira has a periodic variation between third or fourth magnitude and ninth magnitude over an average period of 331 days.

- It can sometimes reach second magnitude and once, in 1779, almost attained first magnitude.

- At times Mira is easily visible to the naked eye, being the brightest star in that part of the sky, while at others it cannot be seen without a **telescope**.

Mira variables

Reddish **giant** stars that change in brightness over a wide range – as much as 10 **magnitudes** – over periods ranging from 100 to 500 days.
See also: *Mira.*

mirror

In all large modern astronomical **telescopes** the image is formed by a parabolic mirror. Reflecting telescopes have been favoured by astronomers over refractors, which are telescopes using lenses. Mirrors do not suffer from **chromatic aberration** – the coloured fringes around images produced by lenses – and can be supported all over, whereas lenses can be supported only at their edges, and can become distorted. The mirror usually consists of a piece of glass, honeycombed to combine lightness with strength, coated with a film of metal deposited in a vacuum. Secondary mirrors direct the light reflected by the objective, or main, mirror to where it is needed. Advanced modern telescopes use adaptive optics, in which rapidly vibrating pistons constantly alter the shape of the mirror to compensate for the distortion of the image caused by the shifting of the Earth's **atmosphere**.

See also: *Gemini Telescopes, James Clerk Maxwell Telescope, Keck Telescopes.*

missing mass

The difference between the amount of matter in the universe that can be observed by the radiation it gives out and the amount that is probably present, judging by its gravitational effects and some theories of the origin of the universe. The observable matter is 10% or less of the amount that theorists believe is present. The missing mass may be in the form of difficult-to-observe objects like **brown dwarfs**, and cold intergalactic gas; but it may also consist of a sea of particles called WIMPs (weakly interacting massive particles), not yet detected in the laboratory,

See also: *dark matter.*

MMT

Telescope with 6.5-m/21.3-ft **mirror**, located at Mt Hopkins, Arizona, operated by the University of Arizona and the Smithsonian Institute. Its name is derived from 'Multiple-Mirror Telescope', because the instrument began life with a unique six-part mirror, each part of 1.8-m/72-in aperture. The combination was equivalent to a single 4.5-m/176-in mirror. It was built in parts because the shapes of the six smaller pieces could be better controlled than that of a single large one, and modern technology makes it possible to combine their images accurately. But the design was abandoned in favour of the single mirror that replaced it.

Moho

Contraction of Mohorovicic discontinuity, the boundary between the Earth's crust and the underlying mantle. It lies about 10–12 km/6–7 mi beneath the oceans and about 33–35 km/20–22 mi beneath the **continents**. Its existence was revealed when seismic records showed that earthquake shock waves travel faster beneath these depths.

Monoceros
A **constellation** on the celestial equator (see **celestial sphere**), represented as a unicorn. Although it includes no bright stars, it lies in the **Milky Way** and contains many faint stars, star clusters and **nebulae**.

moon
A natural **satellite** of a **planet**. The four gas giants of the outer **Solar System** have scores of moons altogether; the rocky terrestrial planets of the inner Solar System have only a few: the Moon, circling the Earth, and **Phobos** and **Deimos** circling **Mars**. Most of the moons of the giant planets formed with the parent planets early in the history of the Solar System, but a few of the outermost ones were formed separately and later captured.

MOONS CIRCLING PLANETS WITHIN OUR GALAXY

Planet	No. of moons	Planet	No. of moons
Mercury	0	Saturn	19
Venus	0	Uranus	17
Earth	1	Neptune	8
Mars	2	Pluto	1
Jupiter	16		

Moon
The natural **satellite** of the Earth. Though its mass is only one-eightieth of the Earth's it is one of the largest satellites in the **Solar System** in relation to its parent **planet**. It **orbit**s in a west-to-east direction every 27.32 days in relation to the stars (the sidereal month). It rotates on its axis in the same time, so it keeps one side permanently turned towards Earth. The Moon has no **atmosphere** and was thought to have no water till ice was discovered on its surface in 1998. On its sunlit side, temperatures reach

Moon *A view of the farside of the Moon's surface taken from Apollo 11.*

110°C/230°F, but during the two-week lunar night the surface temperature drops to –170°C/–274°F.

The phases of the Moon
The Moon has no light of its own; it shines by reflected sunlight. It goes through a cycle of phases as different amounts of the illuminated side are presented to the Earth. It waxes (grows) from new (dark) via first quarter (half Moon) to full, and wanes (decreases) again to new every 29.53 days. This is the **synodic** month, also known as a lunation.

MOON MEMORY AID

To remember the order of the phases of the Moon (to tell whether the current appearance of the Moon indicates waxing or waning), remember:

What's up, DOC?

D indicates a waxing half-Moon (as seen from the Earth's northern hemisphere) with the curve on the right, **O** indicates a full Moon, and **C** indicates a waning Moon with the curve on the left, heading towards a new Moon.

Origins
The origin of the Moon is still open to debate. Scientists have suggested the following theories: that it split from the Earth; that it was a separate body captured by Earth's **gravity**; that it formed in orbit around Earth; or that it was formed from debris thrown off when a body the size of **Mars** struck Earth.

MOON RESEARCH
- The far side of the Moon was first photographed from the Soviet *Lunik 3* in October 1959. Much of our information about the Moon has been derived from these photographs and from other photographs and measurements taken by US and Soviet Moon probes.
- Geological samples have been brought back by US Apollo astronauts and by Soviet Luna robot probes.
- Experiments set up by the US astronauts in 1969–72 continued to provide data after the astronauts had returned to Earth. They included seismic readings and Earth–Moon distance measurements.
- The US probe Lunar Prospector, launched in January 1998, examined the composition of the lunar crust, recorded **gamma rays**, and mapped the lunar **magnetic field**.

Structure and surface features

- The Moon's composition is rocky, with a surface heavily scarred by **meteoroid** impacts that have formed craters up to 240 km/150 mi across.

- Seismic observations indicate that the Moon's surface extends downwards for tens of kilometres; below this crust is a solid mantle about 1,100 km/690 mi thick, and below that a silicate core, part of which may be molten.

- Rocks brought back by astronauts show the Moon is 4.6 billion years old, the same age as Earth.

- The Moon is made up of the same chemical elements as Earth, but in different proportions, and differs from Earth in that most of the Moon's surface features were formed within the first billion years of its history when it was hit repeatedly by meteoroids. In the absence of an atmosphere and with very little geological activity, these have not been destroyed or weathered as features on the Earth have.

- The youngest craters are surrounded by bright rays of ejected rock.

- The largest scars have been filled by dark lava to produce the lowland plains called seas, or maria (plural of '**mare**'). These dark patches form the 'man-in-the-Moon' pattern.

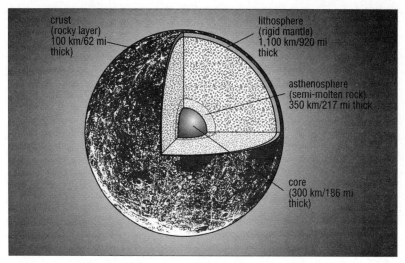

Moon *The Moon has a small core surrounded by a layer of partially molten rock, a thick solid mantle surrounded by a thin rocky crust.*

- Inside some craters that are permanently in shadow is 300 million to 300 billion tonnes of ice, existing as a thin layer of crystals.

- One of the easiest features to observe on the Moon is the crater Plato, which is about 100 km/62 mi in diameter and 2,700 m/8,860 ft deep, and at times is visible with the naked eye alone.

- The US lunar probe *Clementine* discovered an enormous crater on the far side of the Moon, near the south pole, in 1994. The crater is 2,500 km/1,563 mi across and 13 km/8 mi deep, making it the largest known crater in the Solar System.

MOON: STATISTICS

Average distance from Earth	Orbital period (years)	Equatorial diameter (km/mi)	Mass (Earth = 1)
384,400 km/238,855 mi	27.32 days	29.53 days	3,476 km /2,160 mi

Density (water = 1)	Gravity (Earth = 1)	Escape velocity	
3.34	0.16	2.38 kps/1.48 mps	

Mount Palomar

Astronomical **observatory**, 80 km/50 mi north-east of San Diego, California. It has a 5-m/200-in diameter **telescope** called the Hale reflector. Completed in 1948, Mount Palomar was the world's premier observatory during the 1950s.

Mount Wilson

Site near Los Angeles, California, of the 2.5-m/100-in Hooker **telescope**, opened in 1917, with which Edwin **Hubble** discovered the expansion of the universe. Two solar telescopes in towers 18.3 m/60 ft and 45.7 m/150 ft tall, and a 1.5-m/ 60-in reflector, opened in 1908, also operate there.

Mullard Radio Astronomy Observatory

Radio **observatory** of the University of Cambridge, UK. Its main instrument is the Ryle Telescope, consisting of eight dishes 12.8 m/42 ft wide, moving on a track 5 km/3 mi long, opened in 1972.

nadir

The point on the **celestial sphere** vertically below the observer and hence diametrically opposite the **zenith**.

NASA

The National Aeronautics and Space Administration, the US government agency for spaceflight and aeronautical research. It was founded in 1958 by the National Aeronautics and Space Act. Its headquarters are in Washington, DC, and its main installation is at the Kennedy Space Center at **Cape Canaveral** in Florida. NASA's early planetary and lunar programmes included *Pioneer* spacecraft from 1958, which gathered data for the later crewed missions. The most famous of these took the first men to the Moon in *Apollo 11* on 16–24 July 1969. NASA spacecraft have also visited or flown by every **planet** in the **Solar System** except **Pluto**. In the early 1990s, NASA moved towards lower-budget `Discovery missions´, which should not exceed a budget of \$150 million (excluding launch costs), and should not have a development period of more than three years.
See also: *European Space Agency.*

navigation

The science and technology of finding the position, course, and distance travelled by a ship, plane, or other craft. Traditional methods depended heavily on astronomical observations using the sextant, which could measure the **altitude** of the Sun, the Moon or a star, and accurate **time measurement**. Later, highly sophisticated electronic methods, employing radio beacons, displaced these methods. Today instant and accurate readings of position can be obtained from signals received from GPS (Global Positioning System) **satellites**, introduced in 1992.

POSITION FROM THE SKY

▪ GPS features a 'constellation' of satellites that enable civilian users (including motorists and walkers) to fix their position (from any three or more satellites) to within 15 m/50 ft. The US military and its allies can use GPS to get even better accuracy.

▪ In 1992, 85 nations agreed to take part in trials of a new navigation system which makes use of surplus military space technology left over from the Cold War. The new system, known as FANS, or Future Navigation System, will make use of Russian *Glonass* satellites as well as the US GPS satellites.

▪ FANS will be used in conjunction with four *Inmarsat* satellites to provide worldwide communications between pilots and air-traffic controllers.

NEAR

NASA's Near-Earth Asteroid Rendezvous mission, launched in February 1996 to study the asteroid **Eros**. It aimed to be the first space probe to go into orbit around a small body. It would ascertain what asteroids are made of, how their composition compares with that of **meteoroids** and what they tell us of the origin of the **Solar System**.

In 1997 NEAR flew past another **asteroid**, Mathilde, revealing a 25-km/15.5-mi **crater** dominating the tiny world, which is only 53 km/33 mi across. On 20 December 1998 an attempt to rendezvous with Eros failed when a thruster misfired. Communications were regained after 27 hours of silence, and rendezvous was rescheduled for February 2000. The following yearlong encounter was designed to permit study of the asteroid's composition from as close as 50 km/30 mi, using instruments including the onboard **X-Ray/Gamma-Ray** spectrometer and Near-Infrared Spectrometer.

nebula

A cloud of gas and **dust** in space. Vast, cool nebulae are the birthplaces of stars. **Planetary nebulae** consist of hot globes of gas thrown off from dying stars; other nebulae are the remnants of **supernovae**. Formerly the term included **galaxies**.

• Early in the 18th century astronomers adopted the word nebula, derived from the Latin for 'foggy' or 'misty', to describe any non-cometary celestial object that appeared hazy in the **telescopes** then in use.

• The first list of them was drawn up by Charles Messier (1730–1817) in 1784. The **Messier catalogue** numbered 103 prominent objects in the northern sky.

- The nebulae listed in the Messier Catalogue include: M1, the remains of a supernova, known as the **Crab Nebula**; M42, the large gaseous **Orion Nebula**; and M57, a planetary nebula in **Lyra**, known as the Ring nebula.

- As **telescopes** improved, many of the objects previously considered nebulous were seen to be collections of separate stars, and the belief began to grow that all such objects would in time be resolved in this way.

- This belief was shattered in 1864 when William Huggins (1824–1910) observed that NGC 6543, a bright planetary nebula in the constellation **Draco**, did not have a continuous **spectrum** like that of a star, but one consisting of bright lines characteristic of a glowing gas at very low density. The Orion nebula was found to have a similar spectrum.

- It was realized that the gaseous nebulae with this kind of spectrum were of two main types: the planetary nebulae (small, compact, with clear-cut edges) and the diffuse nebulae (very much more extensive, cloudlike in their structure, with no clearly defined boundaries).

HEAVENLY ELEMENT

Some of the spectral lines in nebular **spectra** were clearly due to **hydrogen**, but the origin of the strongest ones could not at first be established. They were therefore attributed to a hypothetical element called 'nebulium', just as strong lines in the solar **spectrum** had been (correctly) attributed to an unknown element, which was christened **helium** before it was identified on **Earth**. It was not until 1927 that the nebulium lines were recognized as being produced by known elements under the extreme conditions of interstellar space.

Types of nebulae
Nebulae are classified according to whether they emit, reflect, or absorb **light**.

- An *emission nebula*, such as the Orion nebula, glows brightly because its gas is energized by stars that have formed within it.

- In a *reflection nebula*, starlight reflects off grains of dust in the nebula, such as surround the stars of the Pleiades cluster.

- A *dark nebula* is a dense cloud, composed of molecular hydrogen, which partially or completely absorbs light behind it. Examples include the Coalsack nebula in **Crux** and the Horsehead nebula in **Orion**.

Neptune

The eighth **planet** in average distance from the Sun. It is a gas giant. At its centre there is believed to a central rocky core covered by a layer of ice. Neptune is not visible to the naked eye. Its existence was predicted by a French mathematician, Urbain Leverrier (1811–1877), from disturbances in the movement of **Uranus**. An English astronomer, John Couch Adams (1819–1892), independently calculated the planet's position.

Neptune's atmosphere

Neptune's **atmosphere** contains methane, which absorbs red light and gives the **planet** a blue colouring. The atmosphere consists primarily of **hydrogen** (85%), **helium** (13%) and methane (1–2%). Neptune has the highest winds in the **Solar System**. *Voyager 2*, which passed **Neptune** in August 1989, revealed various cloud features, notably an Earth-sized oval storm cloud, the Great Dark Spot. This was similar to the Great Red Spot on **Jupiter**, but images taken by the **Hubble Space Telescope** in 1994 show that the Great Dark Spot has disappeared. A smaller dark spot, DS2, has also gone.

NEPTUNE'S SATELLITES

- Of Neptune's eight **moons**, two (**Triton** and Nereid) are visible from **Earth** in large **telescopes.**
- Nereid is the outermost moon of Neptune. It was discovered in 1949 by the Dutch-born US astronomer Gerard Kuiper.
- Nereid is 340 km/210 mi in diameter and orbits Neptune in a highly elliptical **orbit**, ranging from 1.4 million km/0.87 million mi to 9.6 million km/6 million mi from the planet. It has a **period** (the time it takes to circle the planet) of 360 days.
- Six **satellites** were discovered by the *Voyager 2* probe in 1989.
- One of these, Proteus (diameter 415 km/260 mi), is, surprisingly, larger than Nereid (300 km/200 mi).

NEPTUNE MNEMONIC

The initials of the names of the moons of Neptune, in order outwards from the planet, are the initials of the words in:

Neptune's tiny dancing girls look pretty to-night

that is:

Naiad, Thalassa, Despina, Galatea, Larissa, Proteus, Triton, Nereid

Neptune's rings
There are four faint rings, named after astronomers: Galle, Leverrier, Arago, and Adams (in order from the planet outwards). Galle is the widest at 1,700 km/1,060 mi. Leverrier and Arago are divided by a wide diffuse particle band called the 'Plateau'.

NEPTUNE: STATISTICS

Average distance from Sun	Average distance from Sun (Earth = 1)	Orbital period (years)	Equatorial diameter	Mass (Earth = 1)	Tilt of equator
4.4 billion km/ 2,794 billion mi	30.06	164.8	48,600 km/ 30,200 mi	17.2	28.8°

Density (water = 1)	Escape velocity	Rotation period	Known satellites
2.3	25 kps/15.5 mps	16 hr 7 min	8

neutrino

Any of three types of subatomic particles (and their antiparticles). Neutrinos are uncharged leptons. They probably have mass, though it is very small. The three types of neutrino are the electron-neutrino, the muon-neutrino, and the tau-neutrino.

- Unimaginably vast numbers of neutrinos swarm through space, and through the Earth and our bodies. They react so rarely with other kinds of matter that a neutrino can travel through light years of lead and have only a 50% chance of interacting with another subatomic particle on the way.

- **Supernova** 1987A was the first object outside the **Solar System** to be observed by neutrino emission.

- The Sun emits neutrinos, but in smaller numbers than theoretically expected. The shortage of solar neutrinos is one of the biggest mysteries in modern astrophysics.

- If neutrinos have a mass, the fact that there are such huge numbers means that they will have a major gravitational effect in slowing the expansion of the universe, and they will account for much of the **missing mass** in the universe.

neutrino astronomy

Using **neutrinos** to learn about events in distant space. The surface of the Earth is constantly bombarded by neutrinos from the cosmos, but they are extremely difficult to detect. A neutrino telescope consists of a huge tank of liquid, deep underground so that it is shielded from all other types of **cosmic rays** and natural radioactivity by the surrounding rock. Such detectors have measured numbers of neutrinos coming from the Sun and found that they are puzzlingly fewer than theoretically predicted. They have also detected bursts of neutrinos from **supernova** explosions. It may be possible to use beams of artificially produced neutrinos to '**X-ray**' the Earth.

neutron

An uncharged subatomic particle found in the heart of all atoms except those of ordinary **hydrogen**. They play a key role in the nuclear processes that provide the heat and light of the Sun and stars.

- The neutron is made up of three quarks.

- Neutrons have about the same mass as **protons**, about 2,000 times those of **electrons**.

- They contribute to the mass of atoms but have almost no effect on their chemistry. The **isotopes** of a single element differ only in the number of neutrons in their nuclei but have identical chemical properties.

- Outside a nucleus, a free neutron is unstable, decaying with a half-life of 11.6 minutes into a proton, an electron, and an antineutrino (see **neutrino**). The process by which a neutron changes into a proton is called beta decay.

- The neutron was discovered by the British chemist James Chadwick (1891–1974) in 1932.

    See also: *carbon–nitrogen cycle, neutron star, proton–proton cycle.*

neutron star

A very small, 'superdense' star composed mostly of **neutrons**. Neutron stars are thought to form when massive stars explode as supernovae.

- When a **supernova** explodes, the **protons** and **electrons** in the atoms of the remnant of the star merge, owing to intense gravitational collapse, and form neutrons.

- A neutron star may have a mass of up to three times that of the Sun, compressed into a globe only 20 km/12 mi in diameter.

MASSIVELY AMAZING

Neutron stars are so condensed that a fragment the size of a sugar cube would weigh as much as all the people on Earth put together.

- If the mass of the supernova remnant is any greater, its **gravity** is so strong that it shrinks even further to become a **black hole**.

- Being so small, neutron stars can spin very quickly. The rapidly flashing radio sources called **pulsars** are neutron stars. The flashing is caused by a rotating beam of radio energy like a lighthouse beam.

New Technology Telescope (NTT)

Optical **telescope** located at the **European Southern Observatory**, at La Silla, Chile. The NTT came into operation in 1991. It has a thin, lightweight **mirror**, 3.38 m/11.75 ft across, which is kept in shape by computer-adjustable supports to produce a sharper image than is possible with conventional mirrors. Such a system is termed active optics.

Newton, Isaac (1642–1727)

English physicist and mathematician who laid the foundations of physics as a modern discipline. The achievements with greatest implications for astronomy are his mechanics, his theory of universal **gravity**, and his analysis of **light** into the **spectrum**.

Astronomical work

- Newton developed the three principles now called 'Newton's laws of motion'.

- Newton discovered the universal law of gravitation. Together with his laws of motion, it successfully explains most of what is known about the movement of bodies in the **Solar System**.

- In 1679 Newton successfully calculated the Moon's motion on the basis of his theory of gravity.

- He also found that his theory of gravity explained **Kepler's laws** of planetary motion, which had been derived by the German astronomer Johannes **Kepler** on the basis of observations of the **planets**.

- His optical experiments with **prisms** were the basis of the 19th-century development of **spectroscopy**.

- He invented the reflecting **telescope**.
  See also: *relativity.*

nova

A star that suddenly increases in brightness by 10,000 times or more, remains bright for a few days, and then fades away and is not seen again for many years, if at all.

- Although the name comes from the Latin 'new', photographic records show that such stars are not really new, but are already existing faint stars undergoing an outburst of energy that temporarily gives them an absolute **magnitude** in the range –6 to –10, at least 100,000 times brighter than the Sun.

- Two or three novae are detected in our **Galaxy** each year, but on average one is sufficiently close to us to become a conspicuous naked-eye object only about once in ten years.

- Novae very similar to those appearing in our own Galaxy have also been observed in other **galaxies**.

Notable novae

Novae are named according to the **constellation** and year in which they appear. Bright ones in the present century that have been intensively studied spectroscopically include:

- Nova Persei 1901
- Nova Geminorum 1912
- Nova Aquilae 1918 (which became almost as bright as **Sirius**)
- Nova Pictoris 1925
- Nova Herculis 1934
- Nova Puppis 1942.

The nova phenomenon

The rate of increase in brightness, the maximum brightness reached, and the rate of fading differ from nova to nova, but spectroscopic observations indicate that the phenomenon in each case is essentially the same. The star blows off its outer shell with velocities of ejection of the order of 1,000 kps/625 mps, the total mass of the shell being about 1/10,000–1/100,000 that of the Sun. The growing shell initially behaves like the photosphere of a **supergiant**, and pours out rapidly increasing amounts of white **light**. As the expansion continues, however, it grows more and more diffuse so that conditions in it become more like those in a gaseous **nebula**. The continuous **spectrum** fades, leaving mainly monochromatic (single-wavelength) radiations that fade much more slowly.

BIOGRAPHY OF A NOVA

Nova Cygni 1975 was first seen on 29 August, with a magnitude of 3.0. By 31 August it had attained its maximum brightness of 1.8, which it retained for a day, and it was below naked-eye visibility by 5 September. The star had evidently been fainter than photographic magnitude 21 in 1950, as it does not show on charts. It is visible on photographs taken at the Riga Observatory on 5, 8, and 12 August, only a few days before the outburst, as a star having a magnitude of 16.

THEORY OF THE NOVA

Many novae have been found to be members of close **binary stars**, and it has consequently been hypothesized that the stars subject to a nova outburst are close binaries that have evolved so far that the initially higher-mass member has already become a **white dwarf**, while the other one is in its **red giant** phase. As the outer layers of the red giant swell out, some of the material is attracted to the white dwarf, where the surface gravity is so high that the extra matter produces a sufficient rise in the pressure and temperature to start **proton-proton** reactions in the **hydrogen** still remaining in the outer layers. The energy thus suddenly released ejects the surface layer into space so that the remainder can relapse to its former white-dwarf state. (See **mass transfer in binary systems**.) This theory implies that the same system may suffer several nova outbursts, and this does seem to be the case: there are a number of stars, the so-called recurrent novae, for which more than one such outburst has been observed.

nucleosynthesis

The building up of larger atomic nuclei from smaller ones in nuclear reactions at the birth of the universe in the **Big Bang** and thereafter in the centres of stars.

See also: *carbon–nitrogen cycle, proton–proton cycle.*

nutation

A slight 'nodding' of the Earth in space, caused by the varying gravitational pulls of the Sun and Moon. Nutation changes the angle of the Earth's axial tilt (average 23.5°) by about 9 seconds of arc to either side of its mean position, a complete cycle taking just over 18.5 years.

obliquity of the ecliptic
The angle by which the **ecliptic**, the Sun's yearly path across the **celestial sphere**, is tilted in relation to the celestial equator. It is approximately equal to 23.5° and is responsible for the **seasons**.

observatory
A site or facility for observing scientific phenomena – in the case of **astronomy** specifically, celestial bodies. Observatories may be ground-based, carried on aircraft, or sent into orbit as **satellites**, in **space stations**, and on the **space shuttle**.

- The earliest recorded observatory was built in Alexandria, North Africa, by Ptolemy Soter in about 300 BC.

- The erection of observatories was revived in West Asia about AD 1000, and extended to Europe.

- The pre-telescopic observatory built on the island of Hven (now Ven) in Denmark in 1576 for Tycho **Brahe** was elaborate, but survived only to 1597.

- The modern observatory dates from the use of the **telescope** in astronomy from 1609. In the 17th century observatories were built in Paris in 1667 and in London at Greenwich (the **Royal Greenwich Observatory**) in 1675 and at Kew.

- Most early observatories were near towns, but with the advent of big telescopes, clear skies with little background light became essential and observatories moved to high, remote sites.

Optical observatories
The most powerful optical telescopes covering the sky are at:

- **Mauna Kea,** Hawaii, USA
- **Mount Palomar**, California, USA
- **Kitt Peak** National Observatory, Arizona, USA
- **La Palma**, Canary Islands, Spain
- **Cerro Tololo** Inter-American Observatory, in the Andes

- **European Southern Observatory**, at La Silla and Paranal, Chile
- Siding Spring Mountain, Australia
- Zelenchukskaya in the Caucasus, Russia.

Radio astronomy observatories include:
- **Jodrell Bank**, Cheshire, UK
- **Mullard Radio Astronomy Observatory**, Cambridge, UK
- **Arecibo**, Puerto Rico
- **Effelsberg,** Germany
- Parkes, Australia.

occultation
The hiding of one celestial body by another. The Moon often moves in front of a star or a radio source, making it invisible. The moons of **Jupiter** regularly pass behind the parent **planet**. These occultations should be distinguished from **eclipses** of the **satellites**, when they pass into the shadow of Jupiter and are darkened, though still in our line of sight.

Olbers' paradox
A problem discussed in 1826 by the German physician and astronomer Heinrich Olbers (1758–1840), though others had posed it before him: if the universe is infinite in extent and filled with stars, we would see stars wherever we looked in the sky, and the sky would be as bright as the surface of the Sun. Why, on the contrary, is the sky dark at night? The answer is that **light** travels at a finite speed and the universe is not infinitely old. The greatest extent of the universe that we can see is too small – by an enormous factor – for the stars to fill the sky in the way suggested by Olbers.

Oort cloud
A spherical cloud of inactive **comets** beyond **Pluto**, extending out to about 100,000 **astronomical units** (1.5 light years) from the Sun. The gravitational effect of passing stars and the rest of our **Galaxy** disturbs comets in the cloud so that they fall in towards the Sun on highly elongated **orbits**, becoming visible from Earth. As many as 10 trillion comets may reside in the Oort cloud, named after the Dutch astronomer Jan Oort (1900–1992), who postulated it in 1950.
   See also: *Kuiper Belt.*

open cluster or galactic cluster
A loose cluster of young stars. More than 1,200 open clusters have been catalogued, each containing between a dozen and several thousand stars.

They are of interest to astronomers because they represent samples of stars that have been formed at the same time from similar material. Examples include the Pleiades and the Hyades. Open clusters are gradually dispersed by the gravitational influence of other stars in the **Galaxy**.

See also: *globular cluster.*

Ophiuchus
A large **constellation** along the celestial equator (see **celestial sphere**), known as the serpent-bearer because the constellation Serpens, the Serpent, is wrapped around it. The Sun passes through Ophiuchus each December, but the constellation is not part of the **zodiac**. Ophiuchus contains **Barnard's** star.

open cluster *Photograph of the Pleiades, an open cluster of stars in the constellation Taurus, estimated to contain 300–500 members within a sphere 30 light years across.*

opposition
The moment at which a body in the **Solar System** lies opposite the Sun in the sky as seen from the Earth. Such a body crosses the **meridian** at about midnight. The **inferior planets** cannot come to opposition. It is the best time for observation of the **superior planets** as they can then be seen all night.

orbit
The path of a body moving under the gravitational influence of other bodies in space. The orbits of **planets** around stars, of moons around planets, and of stars around the centres of their **galaxies**, are **ellipses** (which are generally in the shape of flattened circles, though a circle also counts as an ellipse, and orbits can be circular). The path of a spacecraft with its engines turned off, moving freely under gravity, is an orbit; the term is not applied to the path of a spacecraft while its engines are running. (*See illustration on page 141*).

See also: *Kepler's laws.*

Orion
A very prominent **constellation** in the equatorial region of the sky (see **celestial sphere**), identified with the hunter Orion of Greek mythology.

- The bright stars Alpha (**Betelgeuse**), Gamma (**Bellatrix**), Beta (**Rigel**), and Kappa Orionis mark the shoulders and legs of Orion.

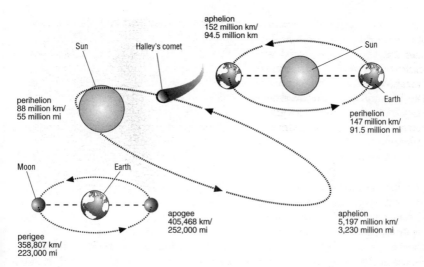

orbit *Most of the orbits of planets, satellites, etc, are elliptical. The nearest point on the orbit of a planet or comet around the Sun is called the perihelion; the farthest point is the aphelion. The nearest point on the orbit of a satellite around a planet is called the perigee; the farthest point is called the apogee.*

- Between them the belt is formed by Delta, Epsilon, and Zeta, three stars of the second **magnitude**, equally spaced in a straight line.

- Beneath the belt is a line of fainter stars marking Orion's sword. One of these, Theta, is not really a star but the brightest part of the **Orion Nebula.**

- Nearby is one of the most distinctive dark **nebulae**, the Horsehead.

- Both Betelgeuse and Rigel are first-magnitude stars and are **supergiants**, but while Betelgeuse is red, Rigel, like most of the other bright stars in the Orion area, is blue.

- The Orionid meteors, which appear about 18–26 October, and seem to radiate from the constellation, are debris from Halley's **comet**.

Orion Nebula

A luminous cloud of gas and **dust** 1,500 **light years** away, in the **constellation Orion**, from which stars are forming. It is about 15 light years in diameter, and contains enough gas to make a cluster of thousands of stars. At the **nebula's** centre is a group of hot young stars, called the Trapezium, which make the surrounding gas glow. The nebula is visible to the naked eye as a misty patch below the belt of Orion.

ozone hole
A thinning in the **ozone layer** that regularly forms over Antarctica as a result of damage to the ozone layer. This has been largely caused by gases produced by industry. These include chlorofluorocarbons (CFCs), but many reactions destroy ozone in the stratosphere: nitric oxide, chlorine, and bromine atoms are implicated.

It is believed that the ozone layer is decreasing at a rate of about 5% every 10 years over northern Europe, with depletion extending south to the Mediterranean and southern USA. However, ozone depletion over the polar regions is the most dramatic manifestation of a general global effect. Ozone levels over the Arctic had by spring 1997 fallen over 10% since 1987, despite the reduction in the concentration of CFCs and other industrial compounds which destroy the ozone when exposed to sunlight. It is expected that an Arctic hole as large as that over Antarctica could remain a threat to the northern hemisphere for several decades.

See also: *ozone layer.*

ozone layer
A thin layer of the gas ozone around 25 km/15 mi high, in the **stratosphere**, which shields the Earth from harmful **ultraviolet radiation**. Ozone is a highly reactive pale-blue gas with a penetrating odour. The ozone molecule is made up of three atoms of oxygen (symbol O_3). It is formed when the molecule of the stable form of oxygen (O_2) is split by ultraviolet radiation or electrical discharge.

See also: *ozone hole.*

Pallas

The second-largest **asteroid** in the **Solar System** (after **Ceres**). It was discovered by the German astronomer Heinrich Olbers (1758–1840) in 1802. It is 522 km/324 mi in diameter. Pallas orbits the Sun at a distance of 414 million km/257 million mi with a **period** (the time it takes to circle the Sun) of 4.61 years.

parallax

The change in the apparent position of an object against its background when viewed from two different positions. In astronomy, nearby stars show a shift owing to parallax when viewed from different positions in the Earth's orbit around the Sun. A star's parallax is used to deduce its distance from the Earth. Nearer bodies such as the Moon, Sun, and **planets** also show a parallax caused by the motion of the Earth. Diurnal parallax is caused by the Earth's rotation.

parallel universes

Universes that are conjectured to exist alongside our own, with no direct communication possible between them. The idea appears in various forms in science-fiction and ideas of parallel universes, or 'multiverses', are flourishing in present-day physics.

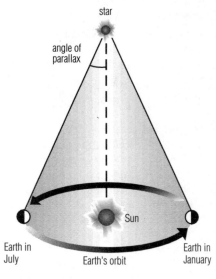

parallax *The parallax of a star, the apparent change of its position during the year, can be used to find the star's distance from the Earth. The star appears to change its position because it is viewed at a different angle in July and January. By measuring the angle of parallax, and knowing the diameter of the Earth's orbit, simple geometry can be used to calculate the distance to the star.*

- It is possible that the entire observable universe is only one of an unimaginably vast number that were created together in the **Big Bang**.

- According to another theory, new universes may be continually born within **black holes**. The **space-time** of these universes would be infinite as far as their inhabitants, if any, were concerned.

- According to the 'many-worlds' interpretation of quantum mechanics, the entire universe is continually branching into a number of universes, each of which branches again, and so on incessantly, each universe following its own history.

parsec
A unit (symbol pc) used for expressing distances to stars and **galaxies**.

- One parsec is equal to 3.2616 **light years**, 2.063 x 10^5 **astronomical units**, and 3.086 x 10^{13} km.

- It is defined as the distance at which a star would have a **parallax** (apparent shift in position) of one second of arc when viewed from two points the same distance apart as the Earth's distance from the Sun.

- This is the same as saying that it is the distance at which one astronomical unit subtends an angle of one second of arc.

partial eclipse
An **eclipse** in which the body being darkened is not wholly obscured. In a **solar eclipse**, as seen from some places the Sun is only partly covered by the Moon; in these places a partial eclipse is seen. In a **lunar eclipse**, the Moon may not completely enter the Earth's shadow but instead slightly to one side of it: this is a partial lunar eclipse.

See also: *annular eclipse.*

Patroclus group
A group of **asteroids** that share the **orbit** of **Jupiter**. The asteroids move around a point 60° behind the **planet**.

See also: *Achilles group; Trojan asteroids.*

Pegasus
A **constellation** of the northern hemisphere, near **Cygnus**, represented as the winged horse of Greek mythology.

- Pegasus is the seventh-largest constellation in the sky.

- Its main feature is a square outlined by four stars, one of which (Alpherat) is actually part of the adjoining constellation **Andromeda**.

- Diagonally across from Alpherat is Markab (or Alpha Pegasi), about 100 **light years** distant from Earth.

penumbra
1 The outer part of a shadow, which receives some but not all of the **light** from the source of illumination. When the Moon passes through the penumbra of the Earth's shadow, the Moon receives some sunlight directly from the Sun, and there is a **partial eclipse**. In a **solar eclipse**, a partial eclipse is seen from those places where the penumbra of the Moon's shadow falls.
2 The outer part of a **sunspot**, not as dark as the central **umbra**.

periastron
The point at which an object travelling in an elliptical **orbit** around a star is at its closest to the star; the point at which it is farthest from the star is known as the apastron.

perigee
The point at which an object, travelling in an elliptical **orbit** around the Earth, is at its closest to the Earth. The point at which it is farthest from the Earth is the **apogee**.

perihelion
The point at which an object, travelling in an elliptical **orbit** around the Sun, is at its closest to the Sun. The point at which it is farthest from the Sun is the **aphelion**.

period
In general, the time it takes any repetitive process to go through one complete cycle. In astronomy:
1 The time it takes a celestial object to revolve around another one, or to rotate once on its own axis. The object's **sidereal period** is the time taken with reference to a background star. The sidereal period of a **planet** is its 'year'. The object's **synodic period** is the time taken to complete a full circle as seen from Earth.
2 The time it takes a **variable** star to go through one cycle, from maximum brightness to maximum brightness.

Perseus
A bright **constellation** of the northern hemisphere, near **Cassiopeia**. It is represented as the mythological hero Perseus. The head of the decapitated Gorgon, Medusa, is held by Perseus, its eye marked by **Algol** (Beta Persei).

- Algol is the best known of the eclipsing **binary** stars.
- Perseus lies in the **Milky Way** and contains the Double Cluster, a twin **globular cluster** of stars called h Persei and Chi Persei. They are just visible to the naked eye as two hazy patches of light close to one another.
- Every August the Perseid **meteor shower** radiates from the constellation's northern part. The Perseids are remnants of **Comet** 1862 III.

phase

The apparent shape of the illuminated part of the Moon or a **planet** as seen from the Earth.

- The Moon is gibbous (more than half but less than fully illuminated) when between first quarter and full, or between full and last quarter.
- **Mars** can appear gibbous at quadrature (when it is at right angles to the Sun in the sky).
- The gibbous appearance of **Jupiter** is barely noticeable.
- The **inferior** planets, whose orbits lie within that of the Earth, can also undergo a full cycle of phases, as can an **asteroid** passing inside the Earth's **orbit**.

THE MOON'S PHASES

- new (Moon between the Earth and the Sun)
- first quarter (Moon at 90° eastern elongation from the Sun)
- full (Moon opposite the Sun)
- last quarter (Moon at 90° western elongation from the Sun).

Phobos

The larger of the two moons of **Mars**. It was discovered in 1877 by the US astronomer Asaph Hall (1829–1907). It is an irregularly shaped lump of rock, cratered by **meteoroid** impacts. It is thought to be an **asteroid** captured by the gravity of Mars.

PHOBOS: STATISTICS

Diameter	Distance from centre of planet	Period
27 x 22 x 19 km/ 17 x 13 x 12 mi	9,400 km/5,840 mi	0.32 days

photosphere

The visible surface of the Sun, which emits light and heat. About 300 km/ 200 mi deep, it consists of incandescent gas at a temperature of 5,530°C/9,980°F. Rising cells of hot gas produce a mottling of the photosphere known as **granulation**, each granule being about 1,000 km/620 mi in diameter. The photosphere is often marked by large, dark patches called **sunspots**.

Pic du Midi Observatory

Astronomical **observatory** on a mountain in the French Pyrenees. The observatory was established in 1878 at the high altitude of 2,865 m/9,400 ft to avoid the haze of the atmosphere at lower altitudes. There are a number of reflecting **telescopes** there, the largest being a 2-m/79-in instrument.

Pioneer probe

Any of a series of US **Solar System** space probes that were active in the period 1958–78.

Pioneer probe *Artist's view of Pioneer probe over Jupiter's Red Spot.*

Pioneer missions

Mission number	Notes
1, 2, 3	Launched in 1958, these were intended Moon probes, but *Pioneer 2*'s launch failed, and *1* and *3* failed to reach their target, although they did measure the Van Allen **radiation belts.**
4–9	Went into solar **orbit** to monitor the Sun's activity during the 1960s and early 1970s.
10	Launched in March 1972, was the first probe to reach **Jupiter** (December 1973) and to leave the Solar System (1983). Ceased to operate in April 1997.
11	Launched in April 1973, passed Jupiter (December 1974), and was the first probe to reach **Saturn** (September 1979), before also leaving the Solar System. Ceased to function in 1995. Both *Pioneer 10* and *11* carry plaques containing messages from Earth in case they are found by other civilizations among the stars. *Pioneer Venus Orbiter, Pioneer Venus Probe* launched May and August 1978. *Orbiter* surveyed the **atmosphere** and surface of **Venus**, and was still active in 1991. *Probe* dropped four vehicles into the atmosphere. The orbital vehicle burned up in the atmosphere of Venus in 1992.

Pisces

An inconspicuous **zodiacal constellation**, lying mainly in the northern hemisphere between **Aries** and **Aquarius**, near **Pegasus**. It is represented as two fish tied together by their tails.

- The Circlet, a delicate ring of stars, marks the head of the western fish in Pisces.

- The constellation contains the vernal **equinox**, one of the two points at which the Sun's path around the sky (the **ecliptic**) crosses the celestial equator (see **celestial sphere**).

- The Sun reaches this point around 21 March each year as it passes through Pisces from mid-March to late April.

- In astrology, the dates for Pisces are between about 19 February and 20 March (see **precession**).

Planck time

An inconceivably early moment after the **Big Bang** in which the universe was born, at which the force of **gravity** became distinct from the other fundamental forces and the laws of physics as presently understood began to be applicable. It is equal to 10^{-43} seconds. Today's science can say nothing about times earlier than this. Immediately after this time the universe consisted of a seething mass of matter and energy and was expanding at a colossal rate.

See also: *universe, future of.*

planet

A large celestial body in **orbit** around a star, composed of rock, metal, or gas. There are nine planets in the **Solar System**: **Mercury**, **Venus**, **Earth**, **Mars**, **Jupiter**, **Saturn**, **Neptune**, **Uranus**, and **Pluto**.

- The inner four planets, called the terrestrial planets, are small and rocky, and include the Earth.

- The outer planets, with the exception of Pluto, are called the major planets, and are gas giants, consisting of large balls of gas with interiors of rock and liquid.

- The largest planet is Jupiter, which contains a mass equivalent to about 70% of all the planets combined.

- Planets do not produce light, but reflect the light of their parent star.

- As seen from the Earth, the historic planets (the planets out as far as Saturn) are conspicuous naked-eye objects, moving in looped paths against the stellar background. (The word 'planet' is from a Greek word meaning 'wanderer'.)

• The sizes of these loops, which are an effect caused by the Earth's own motion round the Sun, are inversely proportional to the planet's distance from the Earth.

Planet X
A hypothetical **planet** lying beyond **Pluto**. It was suggested by the US astronomer Percival Lowell (1855–1916) to explain irregularities in the **orbits** of **Uranus** and **Neptune** that gravitational disturbance by Pluto did not account for. Since then, most of the unexplained variations in the orbits of Uranus and Neptune have been explained or shown to be unreal, so there is no need to postulate a large planet beyond Pluto. However, the first members of a swarm of small objects, comprising the **Kuiper Belt**, have been discovered.

planetarium
An optical projection device by means of which the motions of stars and **planets** are reproduced on a domed ceiling representing the sky.

planetary rings
Swarms of small bodies, ranging in size from grains of **dust** to boulders, orbiting a **planet** in a limited region. The rings of **Saturn** are broad and bright, though very thin. The rings of **Jupiter**, **Uranus** and **Neptune** are thin and dark. The rings are controlled by the gravitational attraction of the planet's **satellite**s, which trap the ring particles and create gaps, or empty zones, between and within rings. The rings of Saturn may consist of debris from satellites that have collided in the past, or may represent dust that failed to form into a moon at the time of formation of the Saturn system. Planetary rings are constantly renewed by debris eroded from collisions of bodies within the rings.

planetary nebula
A **nebula** that looks compact and hard-edged in a **telescope**. It consists of a shell of gas thrown off by a star at the end of its life. Planetary nebulae have nothing to do with **planets**. They were named by William **Herschel**, who thought their rounded shape resembled the disc of a planet. After a star such as the Sun has expanded to become a **red giant**, its outer layers are ejected into space to form a planetary nebula, leaving the core as a **white dwarf** at the centre.

planetary system

A system of **planets**, moons, **comets** and **asteroids**, together with other **interplanetary matter**, circling a star. We have detailed knowledge of only one, our own **Solar System,** but planets have been detected circling other stars. By the dawn of the year 2000, 28 probable planets were known, all detected within the previous five years. All of these extrasolar ('beyond the Sun') planets are large, ranging from 40% of the mass of **Jupiter** to 10 times its mass. They reveal themselves by the 'wobble' in the motion of the star caused by the planet's **gravity**. Present theories of the formation of planets suggests that planetary systems should be common in our **Galaxy**.

Planet-finding technology

- No images can be made of extrasolar planets with today's technology. The ones found so far have been detected by the **Doppler effect**. Their gravity makes the parent star wobble slightly, altering the wavelength of the star's light cyclically.

- A movement in the star of only 10 kph/6 mph – jogging speed – can be detected.

- In 2010, **NASA** plans to launch the Terrestrial Planet Finder, a huge space **telescope** with four mirrors. It will be able to analyse light from the **atmospheres** of Earthlike planets, picking it out from the flood of starlight with which it would be mixed.

NEW PLANETS

- In 1995 Italian astronomers detected a new planet around 51 Pegasi in the **constellation Pegasus**. It was named 51 Pegasi B and has a mass 44% of Jupiter's.
- In April 1996 another planet was discovered, orbiting Rho Cancri in the constellation **Cancer**.
- Yet another was found in June 1996, this time orbiting the star Tau Boötis, in the constellation **Boötes**.
- A new planet with an **orbit** more irregular than that of any other was discovered by US astronomers in October 1996. The new planet had 1.6 times the mass of Jupiter and orbits the star 16 Cygni B at a distance varying from 90 million km to 390 million km.
- Since then, new discoveries have come thick and fast. Most of the new planets move in highly elliptical orbits. Theorists have had to work hard explaining how these could arise.

planetoid

Another term for **asteroid**.

planisphere

A device for displaying the positions of the stars for any date and time in the year. It consists of two discs mounted concentrically. The upper disc has an aperture corresponding to the horizon of the observer, and can rotate over the lower disc, which is printed with a map of the sky centred on the north or south celestial pole. In use, the observer aligns the time of day marked around the edge of the upper disc with the date marked around the edge of the lower disc. The aperture then shows which stars are above the horizon.

Pluto

The smallest and (for most of the time) outermost **planet** of the **Solar System**. The existence of Pluto was predicted by calculation by Percival Lowell (1855–1916) and the planet was located by Clyde Tombaugh (1906–1997) in 1930.

- Its highly elliptical **orbit** occasionally takes it within the orbit of **Neptune**, as in 1979–99.

- Pluto rotates the 'wrong ' way, compared with most of the other planets.

- Pluto resembles the moons of the outer planets rather than the planets themselves. It has been suggested that it is really the largest member of the **Kuiper Belt**.

- Pluto has a low density, being composed of rock and 'ices', primarily frozen methane; there is a **polar cap** at Pluto's north pole.

- There is a thin **atmosphere** with small amounts of methane gas.

- Pluto's single moon is called **Charon**, and has half the diameter of the parent planet.

PLUTO: STATISTICS

Average distance from Sun	Average distance from Sun (Earth = 1)	Orbital period (years)	Equatorial diameter	Mass (Earth = 1)	Tilt of equator
5.8 billion km/ 3.6 billion mi	39.4	248.5	2,300 km/ 1,438 mi	0.003	122.5°

Density (water = 1)	Escape velocity	Rotation period	Known satellites
2.03	approx.1.6 kps/ 1 mps	6.39 Earth days	1

Pluto–Kuiper Express

A **NASA** mission to the farthest **planet** in the **Solar System**, **Pluto**, and beyond. Pluto was the only planet not to be visited by spacecraft in the 20th century. The mission plan calls for the launch of a pair of relatively small and cheap spacecraft, weighing less than 100 kg/45 lb each, in 2001. They will pass within 15,000 km/9,000 mi of Pluto and its moon **Charon** around 2013, The mission could include Russian Zond probes to study the thin atmosphere of Pluto. The spacecraft will then travel on to study the **Kuiper Belt**.

polar cap

A region of frozen water, carbon dioxide or other compounds that lies over the polar regions of Earth and **Mars**. Traces of caps can be seen on moons farther out in the **Solar System** and on **Pluto**, though they are less distinct because their surfaces are generally icy. On Earth, the northern polar cap consists of floating ice; the southern cap lies largely on Antarctica. On Mars the caps are predominantly carbon dioxide, with some water-ice. The Martian caps change greatly in size with the Martian seasons. The southern cap is permanent, while the northern one usually disappears in summer.

Polaris or the Pole Star, or the North Star

The bright star closest to the north celestial pole, and the brightest star in the **constellation Ursa Minor**. Its position is indicated by the 'pointers', two stars in **Ursa Major**. It is also known as Alpha Ursae Minoris.

- Polaris is a yellow **supergiant** about 500 **light years** away.

- It currently lies within 1° of the north celestial pole; **precession** (Earth's axial wobble) will bring Polaris closest to the celestial pole (less than 0.5° away) in about AD 2100. Then its distance will start to increase, reaching 1° in 2205 and 47° in 28000.

- Other bright stars that have been, and will again be, close to the north celestial pole are: Alpha Draconis (3000 BC), Gamma Cephei (AD 4000), Alpha Cephei (AD 7000), and **Vega** (AD 14000).

- Polaris is a **Cepheid variable** whose **magnitude** varies between 2.1 and 2.2 over 3.97 days.

Pollux or Beta Geminorum

The brightest star in the **constellation Gemini** and the 17th-brightest star in the night sky.

- Pollux is a yellow star with a true luminosity 45 times that of the Sun.

- It is 35 **light years** away from the Sun.

The first-magnitude Pollux (Beta Geminorum) and the second-magnitude (that is, fainter) **Castor** (Alpha Geminorum) mark the heads of the twins.

- The two stars may have changed their relative brightness since Johannes Bayer (1572–1625) named them, as Alpha is usually assigned to the brightest star in a constellation.

precession
The slow wobble of the Earth on its axis, like the wobble of a spinning-top. The gravitational pulls of the Sun and Moon on the Earth's equatorial bulge cause the Earth's axis to trace out a circle on the sky every 25,800 years.

- The position of the celestial poles (see **celestial sphere**) is constantly changing owing to precession, as are the positions of the **equinoxes** (the points at which the celestial equator intersects the Sun's path around the sky).

- The precession of the equinoxes means that there is a gradual westward drift in the **ecliptic** – the path that the Sun appears to follow – and in the coordinates of objects on the celestial sphere.

- This is why the dates of the astrological signs of the **zodiac** no longer correspond to the times of year when the Sun actually passes through the

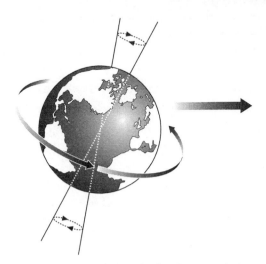

precession *At the present time the north celestial pole points towards the north pole star, Polaris, but in 10,000 years' time the Earth's axis will have moved, because of precession, to point towards Vega, which will become the new 'pole star'.*

constellations. For example, the Sun passes through **Leo** from mid-August to mid-September, but the astrological dates for Leo are between about 23 July and 22 August.

- Precession also occurs in other planets. **Uranus** has the **Solar System's** fastest known precession (264 days).

prime meridian
The line of **longitude** passing through Greenwich, in London, UK. It is taken as the reference **meridian** for longitude and for defining time zones.
   See also: *time measurement.*

prism
In optics, a block of transparent material (plastic, glass, silica), usually with a triangular cross-section, commonly used to split a ray of **light** into its **spectrum**. Prisms are also used as **mirrors** to define the optical path in **binoculars**, camera viewfinders, and periscopes. The colour-dispersing property of prisms was formerly used in **spectroscopy**, but the prism has been largely superseded by the **diffraction** grating, a piece of glass or metal on which is engraved an array of fine parallel lines.

Procyon or Alpha Canis Minor
The brightest star in the **constellation Canis Minor** and the eighth-brightest star in the night sky.

- Procyon is a white star 11.4 light years from the Sun, with a mass of 1.7 Suns.

- It has a **white dwarf** companion that orbits it every 40 years.

- Procyon and **Sirius** are sometimes called 'the dog stars'.

- The name 'Procyon', derived from Greek, means 'before the dog', and reflects the fact that in mid-northern **latitudes** Procyon rises shortly before Sirius.

prominence
A bright cloud of gas that erupts from the Sun into space. Prominences can be 100,000 km/60,000 mi high or more.

- *Quiescent* prominences last for months, and are held in place by **magnetic fields** in the Sun's **corona**.

- *Surge* prominences shoot gas into space at speeds of 1,000 kps/600 mps.

- *Loop* prominences are gases falling back to the Sun's surface after a solar **flare**.

propellant

Either the fuel or the oxidant that a **rocket** carries to drive its engines. The oxidant is a substance that provides oxygen with which the fuel can burn. Powerful propellants include liquid **hydrogen** or liquid kerosene (fuels) and liquid oxygen (oxidant). Propellants can be liquid or solid.

proper motion

The apparent motion of a celestial body across the **celestial sphere** in relation to the background of stars. The proper motion of the Moon is about 0.5° per hour; the proper motion of the fastest-moving star, **Barnard's** Star is about the same distance in 175 years.

protogalaxy

The enormous cloud of gas from which each **galaxy** formed. The hot expanding gas that comprised the newly born universe began to separate into clumps of denser gas about a million years after the **Big Bang**, and these separated to become the protogalaxies. Each protogalaxy in turn began to break up into smaller knots of gas, in which **proto**stars began to form.

proton

A subatomic particle found in the heart of each atom.

- The number of protons in the atomic nucleus determines the chemical identity of the atom.

- The nucleus of the **hydrogen** atom consists of a single proton.

- The nucleus of, for example, iron consists of 26 protons and (usually) 30 **neutrons**.

- The proton has an electric charge equal and opposite to that of the **electron**, and a mass 1,836 times as great.

- The proton's reactions are central to the nuclear processes in stars, in which protons are built up into nuclei of heavier elements.

   See also: *proton–proton cycle*.

proton–proton cycle

The most important process by which **hydrogen** is converted to **helium** by nuclear fusion in the core of the Sun and in stars of similar mass. In heavier stars the **carbon–nitrogen cycle** is more important. In both processes, four **protons** are converted to a helium nucleus, which consists of two protons and two **neutrons**. Positrons, **neutrinos**, and **gamma rays** are emitted. The temperature must exceed several million degrees kelvin for the cycle to start.

protoplanet

The early stage in the formation of a **planet**, when it is a mass of **dust** gathering in the larger disc of gas from which a new **planetary system** is being formed.

protostar

The early stage in the formation of a star, when it has recently condensed out of an interstellar cloud and **hydrogen**-burning has not yet started.

- Protostars are very hot, deriving their energy from gravitational contraction, but are shrouded in dense gas, so that little visible **light** escapes.

- They can be recognized by their intense **infrared radiation**.

- Protostars have violent stellar winds (the equivalent of our Sun's **solar wind**), which eventually blow away the gas and dust surrounding them.

- Many protostars have been detected in active regions of starbirth such as the **Orion Nebula.**

Proxima Centauri

The star closest to the Sun, 4.2 **light years** away. It is a faint red dwarf, visible only with a **telescope**, and is a member of the **Alpha Centauri** triple-star system. It is called Proxima (Latin for 'near') because it is about 0.1 light years closer to us than its two partners.

Ptolemaic system

The Earth-centred model of the universe worked out by **Ptolemy** in the 1st century AD. The book later called *The Almagest* (he called it *Syntaxis*) contains all his works on astronomical themes, the only authoritative works until the time of **Copernicus**.

Ptolemy (*c.* 100–*c.* 170)

Egyptian astronomer and geographer, full name Claudius Ptolemaeus, whose theories dominated European and Arabic **astronomy** for 1,400 years. His *Almagest* developed the description of the movements of the Sun, Moon, **planets** and stars, in relation to the Earth, which is taken to be the centre of the universe.

AT THE CENTRE OF THINGS

■ Ptolemy was probably inspired by the Greek philosopher Plato (c.427 BC –347 BC), who described the universe as an elaborate system of nested spheres.

- Ptolemy began with the assumption that the Earth was a perfect sphere.
- All **planet** orbits were circular, but those of **Mercury** and **Venus**, and possibly **Mars** (Ptolemy was not sure), were epicyclic – that is, the planets orbited a point that itself was orbiting the Earth.
- The sphere of the stars formed a dome with points of light attached, or with light shining through.
- The system was complex. A partial simplification was achieved by **Copernicus**, whose Sun-centred system (1543) was eventually to sweep away the Ptolemaic system.
- But even Copernicus retained circular **orbits** and epicycles (see **Kepler's laws**).

pulsar

Rapidly rotating **neutron** star, which flashes at radio and other wavelengths as it spins, with a **period** ranging from a few seconds to a few thousandths of a second. Pulsars were discovered in 1967 by Jocelyn **Bell-Burnell**, a postgraduate student on a team led by Antony Hewish at the **Mullard Radio Astronomy Observatory**, Cambridge, in the UK. By 2000 over 1,000 pulsars had been discovered.

- Pulsars slow down as they get older, and eventually the flashes fade. In the case of the Crab pulsar the increase in period is 0.000000300 seconds per day, a figure which is consistent with its age of about 900 years.
- The rates of slowing indicate an active life for a pulsar of about 10^7 years.
- Some pulsars give out flashes of visible **light** as well as radio waves. There is one in the **Crab nebula**, the remnant of a **supernova** that was observed in 1054, and another (estimated to be 11,000 years old) in the **constellation** Vela.

INSIDE A PULSAR

- The existence of neutron stars was predicted theoretically by Lev Landau (1908–1968) in 1932. They have approximately the same mass as the Sun but are only about 10 km/6 mi across. Their density is therefore incredibly high.
- At the surface, where the pressure is lower, there is a crystalline crust about 1 km/0.6 mi thick, formed of closely packed iron nuclei.
- In addition to a steady slowing down, some pulsars show small sudden changes of period. These are called glitches and are thought to be caused by 'starquakes' – sudden changes in the shape of the pulsar's crust.

quasar

Any of a class of incredibly energetic objects, lying at vast distances beyond our **Galaxy**. Quasars lie at the centres of **galaxies**, their energy being derived from stars and gas falling into an immense **black hole** at their nucleus. Most quasars are found in **elliptical galaxies**.

- Quasars were discovered in 1963.

- Quasars appear starlike, and their name is a contraction of 'quasi-stellar object' – they are sometimes still called QSOs.

- Their distance is so great that their apparent brightness represents a huge energy output. Each emits more energy than 100 giant galaxies.

- Quasar light shows a large **red-shift**, indicating that they are very distant.

- Some quasars emit radio waves (see **radio astronomy**), which is how they were first identified, but most are radio-quiet.

- The farthest are over 10 billion **light years** away.

radar astronomy

Studying objects in the **Solar System** by bouncing radio waves off them and analysing the 'echoes'. Radar contact with the Moon was first made in 1945 and with **Venus** in 1961. The travel time for radio reflections allows the distances of objects to be determined accurately. Analysis of the **Doppler effect** in the reflected beam reveals the rotation speed of the object observed, and other features of the echo allow the surface to be mapped. The rotation periods of Venus and **Mercury** were first determined by radar. Radar maps of Venus were obtained first by Earth-based radar and subsequently by orbiting space probes, including *Magellan*.

radiant

A point in the sky from which the **meteors** in a **meteor shower** appear to come. The meteors generally appear a long way from the radiant: it is only when their paths are traced back that they are seen to diverge from one point. The shower receives its name from the **constellation** in which the radiant lies – the Leonids from **Leo**, the Geminids from **Gemini**, and so on.

radiation belts

Regions around a **planet** in which high-energy charged particles have been captured by the planet's **magnetic field**.

- There are two zones of charged particles around the Earth, named the Van Allen belts after the US physicist James Van Allen (1914–   ) who discovered them in 1958.

- The atomic particles come from the Earth's upper **atmosphere** and the **solar wind**.

- The inner belt lies 1,000–5,000 km/620–3,100 mi above the equator, and contains **protons** and **electrons**.

- The outer belt lies 15,000–25,000 km/9,300–15,500 mi above the equator, but is lower around the magnetic poles. It contains mostly electrons from the solar wind.

- The Van Allen belts are hazardous to astronauts, and interfere with electronic equipment on **satellites**.

- Sometimes electrons spiral down towards the Earth, noticeably at polar latitudes, where the magnetic field is strongest. When such particles collide with atoms and ions in the **thermosphere**, light is emitted. This is the origin of the glows visible in the sky as the **aurora**.

radio astronomy

The study of radio waves emitted naturally by objects in space. Radio astronomy began in 1932 when the US astronomer Karl Jansky (1905–1950) detected radio waves from the centre of our **Galaxy**, but the subject did not develop until after World War II, when it benefited from the wartime development of radar. Radio astronomers have detected many types of radio source within our Galaxy and beyond. A major goal in the study of any radio source is to identify it, if possible, with an object that can be detected optically. **Radio telescopes** are routinely networked across countries, continents and the globe: by combining the signals they receive, astronomers discern more detail in the objects they study. Radio astronomers have searched for signals from other civilizations in the Galaxy, and continue to do so, so far without success.

See also: *radar astronomy.*

TYPES OF SIGNAL

Radio emission comes from:
- hot gases (thermal radiation) – for example, between the **galaxies**
- **electrons** spiralling in **magnetic fields** (synchrotron radiation) – for example, in **supernova** remnants
- atoms and molecules in space, 'broadcasting' on specific wavelengths (called 'lines') – the most important of these being the **21-cm** line. This is emitted by ordinary, electrically neutral **hydrogen** gas. Hydrogen is by far the most abundant element in the universe, and the discovery of the 21-cm radiation has allowed astronomers to map the spiral structure of the Galaxy.

There are many kinds of radio source in our Galaxy.
- The remains of supernova explosions, such as the **Crab Nebula** and **pulsars**, are strong radio emitters.
- Short-wavelength radio waves have been detected from complex molecules in dense clouds of gas where stars are forming.
- Strong sources of radio waves beyond our Galaxy include **radio galaxies** and **quasars**. They are more numerous far off in the universe and far in

the past (the radio waves we detect left them billions of years ago), a fact that demonstrates that the universe has evolved with time.

- **Cosmic background radiation** originates from less than a million years after the **Big Bang** explosion that marked the birth of the universe.

radio galaxy

A **galaxy** that is a strong source of radio waves. All galaxies, including our own, emit some radio waves, but radio galaxies are up to a million times more powerful.

- In many cases the strongest radio emission comes not from the visible galaxy but from two clouds, invisible through an optical **telescope**, that can extend for millions of **light years** either side of the galaxy.

- This double structure at radio wavelengths is also shown by some **quasars**, suggesting a close relationship between the two types of object.

- In both cases, the source of energy is thought to be a massive **black hole** at the centre.

- Some radio galaxies are thought to result from two galaxies that are colliding or that have recently merged.

radio telescope

An instrument for detecting radio waves from the universe in **radio astronomy**. Some radio telescopes consist of a metal dish that collects and focuses radio waves in the same way that a concave mirror collects and focuses **light** waves. Other radio telescopes are shaped like long troughs, and some consist of simple rod-shaped aerials.

- Radio telescopes are much larger than optical **telescopes**, because the wavelengths they are detecting are much longer than the wavelength of light.

- Even a large single dish such as that at **Jodrell Bank,** Cheshire, UK can see the radio sky in less detail than a small optical telescope sees the visible sky.

- The largest single dish is 305 m/1,000 ft across; it is built into a natural hollow at **Arecibo**, Puerto Rico.

- **Interferometry** is a technique in which the output from two telescopes is combined to give better resolution of detail than with a single dish. Very long baseline interferometry (VLBI) uses radio telescopes spread across countries or even the world to resolve minute details of radio sources.

- The **VLA** (Very Large Array) in New Mexico, USA, consists of 27 dishes arranged in a Y-shape, which simulates the performance of a single dish 27 km/17 mi in diameter.

Ranger missions

Series of unmanned US space probes that investigated the Moon from 1961 to 1965. The first successful US Moon probe was *Ranger 7*, which sent back 4,316 close-up photographs before it hit the Moon on 31 July 1964. They prepared the way, with other US robot probes, for the *Apollo* manned flights.

RATAN

RATAN (Radio Astronomy Telescope of the Russian Academy of Sciences) is located at Zelenchukskaya, in the Caucasus Mountains of Russia. It consists of radio reflectors in a circle of 600 m/2,000 ft diameter.

rays, lunar

Ejecta (pieces of ejected material) from **impact craters** on the Moon often form long bright streaks known as rays, which in some cases can be traced for thousands of kilometres across the lunar surface.

reddening, interstellar

A change in the colour of starlight due to the absorption of shorter (bluer) wavelengths by **interstellar matter**. The phenomenon yields a great deal of information about the composition of gas and **dust** between the stars, while making it harder to reach accurate conclusions about the nature of stars at great distances in our **Galaxy**.

red giant

Any of a class of large, bright stars with cool surfaces. A red giant represents a late stage in the evolution of a star like the Sun.

- A star swells into a red giant as it runs out of **hydrogen** fuel at its centre.

- The star keeps shining by burning heavier elements, such as **helium**, carbon, and silicon.

- Because of more complex nuclear reactions that then occur in the red giant's interior, it eventually becomes gravitationally unstable and begins to collapse and heat up.

- The result is either the explosion of the star as a **supernova**, leaving behind a **neutron** star, or loss of mass by more gradual processes to produce a **white dwarf**.

- Red giants have diameters between 10 and 100 times that of the Sun.

- A red giant's surface temperature is lower than that of the Sun – about 1,700–2,700°C/3,000–5,000°F.

- Although they are cool, red giants are very bright because they are so large. **See also:** *supergiant*.

red-shift

Lengthening of the wavelengths of light from an object as a result of the object's motion away from us. It is an example of the **Doppler effect**.

- The red-shift in light from **galaxies** is evidence that the universe is expanding.

- Lengthening of wavelengths causes the **light** to move or shift towards the red end of the **spectrum**, hence the name.

- The amount of red-shift can be measured by the displacement of lines in an object's **spectrum**.

- A strong gravitational field can also produce a red-shift in light; this is termed gravitational red-shift.

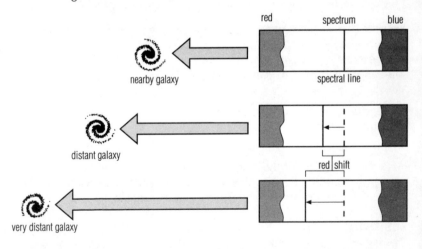

red-shift *The red shift causes lines in the spectra of galaxies to be shifted towards the red end of the spectrum. More distant galaxies have greater red shifts than closer galaxies. The red shift indicates that distant galaxies are moving apart rapidly, as the universe expands.*

reflection nebula

A **nebula** that consists largely of **dust** and shines by reflection of light from nearby stars. One example is the nebula enveloping the Pleiades **open cluster**.

See also: *dark nebulae*.

refraction

The bending of a wave when it passes from one medium into another. The refraction of **light** is important to the observational astronomer. Near the horizon, it causes stars and other celestial objects to appear significantly higher than they really are. The Sun and Moon are noticeably flattened near the horizon because the lower part of the disc is raised slightly more than the top part.

THE BENDING OF LIGHT

Refraction is the effect of the different speeds of wave propagation in two substances that have different densities. The amount of refraction depends on the densities of the media, the angle at which the wave strikes the surface of the second medium, and other factors. Refraction occurs with all types of waves – electromagnetic waves, sound waves, and water waves – and differs from reflection, which involves no change in velocity.

Regulus or Alpha Leonis

The brightest star in the **constellation Leo**, and the 21st brightest star in the sky.

- Regulus is of the first **magnitude**, and has a true **luminosity** 100 times that of the Sun.

- It is 69 **light years** from Earth.

- Regulus was one of the four royal stars of ancient Persia, marking the approximate positions of the Sun at the **equinoxes** and **solstices**. The other three were **Aldebaran**, **Antares**, and **Fomalhaut**.

relativity

The theory that space, time, energy and mass are interdependent and relative to the mutual motion of the observer and the system observed. It was developed by the German-born US physicist Albert **Einstein** in two phases.

- The *special theory of relativity* (1905) started from the assumptions that (1) the laws of nature are the same for all observers in unaccelerated motion and (2) the speed of light is independent of the motion of its source.

- According to the *general theory of relativity* (1915), the geometrical properties of **space-time** were to be conceived of as modified locally by the presence of a body with mass. A planet's orbit around the Sun (as observed in three-dimensional space) arises from its natural trajectory in modified space-time; there is no need to invoke, as Isaac **Newton** did, a force of **gravity** coming from the Sun and acting on the planet.

CONSEQUENCES OF THE SPECIAL THEORY

The special theory has some rather unexpected consequences. Intuitively familiar concepts, like mass, length, and time, had to be modified.

- For example, an object moving rapidly past the observer will appear to be both shorter and heavier than when it is at rest (that is, at rest relative to the observer), and a clock moving rapidly past the observer will appear to be running slower than when it is at rest.
- These predictions of relativity theory seem to be foreign to everyday experience merely because the changes are quite negligible at speeds less than about 1,500 kps/900 mps, and they only become appreciable at speeds approaching the speed of light.
- Einstein showed that, for consistency with the above premises (1) and (2), the principles of dynamics as established by Newton needed modification; the most celebrated new result was the equation $E = mc^2$, which expresses an equivalence between mass (m) and energy (E), c being the speed of light in a vacuum.
- In 'relativistic mechanics', conservation of mass is replaced by the new concept of conservation of 'mass–energy'.

Consequences of the general theory

- Einstein's general theory accounts for a peculiarity in the orbit of the planet **Mercury** that cannot be explained in Newton's theory.

- The new theory also said that light rays should bend when they pass by a massive object. The predicted bending of starlight by the Sun was observed during the **solar eclipse** of the Sun of 1919.

- A third corroboration is found in the shift towards the red in the spectra of the Sun and, in particular, of stars of great density – for example, **white dwarfs** such as the companion of **Sirius**.

- As part of the same effect, time runs slower in a gravitational field than it does in free space. The effect has been confirmed by comparing very accurate clocks in **satellites** with others on the ground.

- General relativity is central to modern astrophysics and cosmology; it predicts, for example, the possibility of **black holes**.

resolving power

1 The ability of a **telescope** (or other optical instrument) to see detail – to distinguish or produce distinct images of small adjacent objects.

2 The ability of a photographic emulsion or other image-producing device to form finely detailed images.

resonance

A large physical effect produced by some relatively small, appropriately timed periodic phenomenon. In everyday life, it is familiar in the sounding of certain strings, and no others, in a piano when a sound is made nearby. The strings that resonate are the ones for which the frequency of the stimulating sound is just right. In astronomy, resonance occurs as a gravitational effect between some orbiting bodies, most noticeably in the existence of gaps in bands or rings of orbiting bodies. Gaps in the rings of **Saturn** mark the locations of resonances with the planet's innermost moons: any object that wandered into such a gap would receive repeated disturbances from the moons, which would force it into another **orbit**. There are gaps in the **asteroid** belt where there are resonances with the giant planet **Jupiter**, orbiting outside the asteroid belt. The very existence of the asteroid belt is probably due to gravitational resonance with Jupiter, which prevented the formation of a full-sized planet at that distance from the Sun.

retrograde

Describing the **orbit** or rotation of a **planet** or **satellite** if the sense of rotation is opposite to the general sense of rotation of the **Solar System**. On the **celestial sphere**, it refers to the temporary motion of planets from east to west against the background of stars, interrupting the normal the normal west-to-east, or *prograde*, motion.

Rigel or Beta Orionis
The brightest star in the **constellation Orion**.

- It is a blue-white **supergiant**, with an estimated diameter 50 times that of the Sun.

- It is 900 **light years** from the Sun.

- Rigel is intrinsically the brightest of the first-**magnitude** stars, its true **luminosity** being about 100,000 times that of the Sun.

- It is the seventh-brightest star in the night sky.

- The name 'Rigel' is derived from the Arabic for 'foot'.

right ascension

The coordinate on the **celestial sphere** that corresponds to **longitude** on the surface of the Earth. It is measured in hours, minutes, and seconds eastwards from the vernal **equinox**, the point where the Sun crosses the celestial equator in the northern spring.

See also: *declination.*

Rigil Kent

Another name for the star **Alpha Centauri**.

Rille

A furrow, trench or narrow valley on **Mars**, or one of the moons of the Solar System.

Roche limit

The distance from a **planet** within which a large moon would be torn apart by the planet's gravitational force, creating a set of rings. The Roche limit lies at approximately 2.5 times the planet's radius (the distance from its centre to its surface).

Roche lobe

The volume of space around one member of a **binary star** in which the star's gravity predominates. The star will capture any matter from its companion that enters its Roche lobe. This can happen if the companion star exhausts its fuel first and swells up to become a **giant** star.

See also: *mass transfer in binary star systems.*

rocket

A projectile or vehicle driven by the reaction of gases produced by a fast-

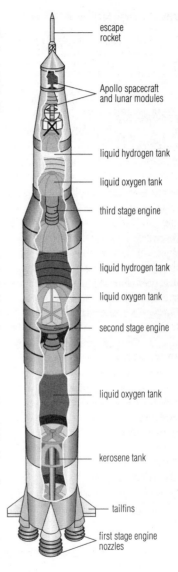

escape rocket

Apollo spacecraft and lunar modules

liquid hydrogen tank

liquid oxygen tank

third stage engine

liquid hydrogen tank

liquid oxygen tank

second stage engine

liquid oxygen tank

kerosene tank

tailfins

first stage engine nozzles

rocket *The three-stage* Saturn V *rocket used in the* Apollo *moonshots of the 1960s and 1970s. It stood 111 m/365 ft high, as tall as a 30-storey skyscraper, weighed 2,700 tonnes/3,000 tons when loaded with fuel, and developed a power equivalent to 50 Boeing 747 jumbo jets.*

burning fuel. Unlike jet engines, which are another type of reaction engine, rockets carry their own oxygen supply to burn their fuel and do not require any surrounding **atmosphere**.

- Rockets have been valued as fireworks for at least 700 years.
- Their intensive development as a means of propulsion to high altitudes, carrying payloads, started only between the two World Wars. The state supported work in Nazi Germany, primarily by Wernher von Braun (1912–1977); Robert Hutchings Goddard (1882–1945) worked in the United States.
- Being the only form of propulsion available that can function in a vacuum, rockets are essential to exploration in outer space.
- Multistage rockets have to be used, consisting of a number of rockets joined together. Each stage is jettisoned as its fuel is exhausted.

Rosetta
A project of the **European Space Agency**, due for launch in 2003, to send a spacecraft to Comet Wirtanen. *Rosetta* is expected to go into **orbit** around the **comet** in 2011 and land two probes on the nucleus a year later. The spacecraft will stay with the comet as it makes its closest approach to the Sun in October 2013.

Royal Observatory, Greenwich
The national astronomical observatory of the United Kingdom, named after its original site at Greenwich, London. The Observatory was founded in 1675 by King Charles II, to provide navigational information for sailors. The eminence of its work resulted in **Greenwich Mean Time** and the Greenwich **meridian** being adopted as international standards of reference in 1884. The Observatory operates **telescopes** on **La Palma** in the Canary Islands, including the 4.2-m/ 165-in William Herschel Telescope, commissioned in 1987.

Ryle Telescope
A **radio telescope** at the **Mullard Radio Astronomy Observatory**, Cambridge, UK. It consists of eight dishes 12.8 m/42 ft wide, strung out in a line along a stretch of railway track 5 km/3 mi long. The signals from the dishes can be combined to provide a highly detailed picture of celestial radio sources. The instrument went into operation in 1972.

Sagittarius

A bright **zodiacal constellation** in the southern hemisphere, represented as a centaur aiming a bow and arrow at neighbouring **Scorpius**, the Scorpion.

- Kaus Australis and Nunki are the brightest stars in Sagittarius.

- The constellation contains many **nebulae**, **globular clusters**, and **open clusters**, and the **Milky Way** passes through it.

- The centre of the **Galaxy** is marked by the radio source Sagittarius A.

- The Sun passes through Sagittarius from mid-December to mid-January. The winter solstice, where the Sun is farthest south of the equator, lies in the constellation.

- In astrology, the dates for Sagittarius are about 22 November–21 December (see **precession**).

Salyut

A series of seven **space stations** launched by the Soviet Union from 1971 to 1982. The name is the Russian for 'salute'. Crews aboard the *Salyut* space stations observed the Earth and space, and carried out the processing of materials in weightlessness. In 1986 the *Salyut* series was superseded by *Mir*, an improved design capable of being enlarged by additional modules sent up from Earth.

- Each *Salyut* was cylindrical in shape and 15 m/50 ft long, and weighed 19 tonnes/21 tons.

- It housed two or three cosmonauts at a time, for missions lasting up to eight months.

Salyut missions

Mission	Notes
1	Launched on 19 April 1971. It was occupied for 23 days in June 1971 by a crew of three, who died during their return to Earth when their Soyuz ferry craft depressurized.
2	In 1973 *Salyut 2* broke up in orbit before occupation.
3	The first fully successful *Salyut* mission was a 14-day visit to *Salyut 3* in July 1974.

Salyut **missions** (*continued*)

Mission	Notes
7	In 1984–85 a team of three cosmonauts endured a record 237-day flight in *Salyut 7*. The craft, the last in the series, crashed to Earth in February 1991, scattering debris in Argentina.

saros cycle

A period of 6,585.3 days, or 18 years and 9 to 11 days (depending on how many leap years there are in the period), after which a **solar** or **lunar eclipse** will be repeated. During this period approximately 29 lunar eclipses and 41 solar eclipses will take place. Eventually the Sun and Moon get out of step and a sequence of saros cycles ends. New ones begin from time. The saros cycle has been known since Babylonian times.

satellite, artificial

An artificial body that permanently orbits a **planet**, moon, or other celestial body. Artificial satellites are used for scientific purposes, communications, weather forecasting, and military applications.

- The first artificial satellite, ***Sputnik 1***, was launched into **orbit** around the Earth by the USSR in 1957.

- The brightest artificial satellites can be seen by the naked eye. Times of appearance are published in some newspapers.

- At any time, there are several thousand artificial satellites orbiting the Earth, including several hundred active satellites, together with satellites that have ended their working lives, and discarded sections of **rockets**.

- Artificial satellites eventually re-enter the Earth's **atmosphere**. Usually they burn up by friction, but sometimes debris falls to the Earth's surface, as with *Skylab* and *Salyut 7*.

Saturn

The second-largest **planet** in the **Solar System**, sixth from the Sun, encircled by bright equatorial rings, easily visible through **binoculars** or a small **telescope**. To the naked eye, Saturn looks yellowish. Viewed through a telescope, it is ochre. Its polar diameter is smaller than its equatorial diameter by 12,000 km/7,450 mi. This is a result of its fast rotation and low density, the lowest of any planet. Its mass is over 90 times that of the Earth, and its **magnetic field** 1,000 times stronger. Saturn is believed to have a small core of rock and iron, encased in ice.

SATURN'S ATMOSPHERE

The visible surface consists of swirling clouds, probably made of frozen ammonia at a temperature of −170 °C/−274 °F. The markings in the clouds are not as prominent as **Jupiter's**. The space probes *Voyager 1* and *Voyager 2* found winds reaching 1,800 kph/1,100 mph

Saturn's satellites

- Saturn has 18 known moons, more than any other planet. There have been reports of possibly 10 other **satellites**, but they have not been confirmed.

- The largest moon, **Titan**, has a dense atmosphere.

- At least seven of the satellites move within the faint, thin outermost rings.

- The **gravity** of the satellites creates the finely detailed structure of the rings.

- Tiny Janus and Epimetheus share the same orbit and continually swap places.

- The outermost satellite, Phoebe, moves in a **retrograde** orbit – that is, the 'wrong' way, opposite to all the other satellites. The orbit is also a highly elongated **ellipse**.

Saturn *Saturn taken from Voyager 1.*

The Voyager probes showed that the rings actually consist of thousands of closely spaced 'ringlets', looking like the grooves in a gramophone record.

- Each ringlet of the rings is made of a swarm of icy particles like snowballs, a few centimetres to a few metres in diameter.

- Outside the A ring is the narrow and faint F ring, which the Voyagers showed to be twisted or braided.

SATURN'S RINGS

The rings of Saturn could be the remains of a shattered moon, or they may always have existed in their present form. From Earth, they appear to be divided into three main sections:

- Ring A is the outermost.
- Ring A is separated from ring B by the **Cassini division,** named after its discoverer, the Italian astronomer Giovanni Cassini (1625–1712). The Cassini Division is 3,000 km/2,000 mi wide.
- Ring B, the middle ring, is the brightest.
- The inner ring C is transparent, and is also called the Crepe Ring.
- The rings visible from **Earth** begin about 14,000 km/9,000 mi from the planet's cloud tops and extend out to about 76,000 km/47,000 mi.
- They are 275,000 km/170,000 mi rim to rim, but only 100 m/300 ft thick.

SATURN: STATISTICS

Average distance from Sun	Average distance from Sun (Earth = 1)	Orbital period (years)	Equatorial diameter	Mass (Earth = 1)	Tilt of equator
1,427 billion km/ 0.886 billion mi	9.5	29,46	120,000 km/ 75,000 mi	95.1	26.75°

Density (water = 1)	Escape velocity	Rotation period	Known satellites
0.69	37 kps/23 mps	10 hr 14 min (equator) 10 hr 40 min (higher latitudes)	18 confirmed

Saturn rocket

A family of large US rockets, developed by German rocket engineer Wernher von Braun (1912–1977) for the *Apollo* project. The two-stage *Saturn IB* was used for launching **Apollo** spacecraft into orbit around the Earth. The three-stage *Saturn V* sent *Apollo* spacecraft to the Moon, and launched the *Skylab* **space station**. The liftoff thrust of a *Saturn V* was 3,500 tonnes. After *Apollo* and **Skylab**, the Saturn rockets were retired in favour of the **space shuttle**.

scintillation

The twinkling of stars caused by the Earth's shifting **atmosphere**, and the fluctuations in radio signals from celestial sources caused by movements in the matter lying along the line of sight.

Adaptive optics

- Although stars twinkle, **planets** do not. Stars are points of light (even in powerful telescopes), while planets have a disc (even though this isn't apparent to the naked eye). The scintillations are averaged out over a planet's disc and the total effect is of a steady brightness.

- Advanced modern **telescopes** using 'adaptive optics' can 'switch off' the twinkling of stars. The shape of the **mirror** is constantly altered by a large number of computer-controlled pistons to cancel out the scintillation and form a sharper image.

Scorpius

A bright **zodiacal constellation** in the southern hemisphere between **Libra** and **Sagittarius**, represented as a scorpion.

- The Sun passes briefly through Scorpius in the last week of November.

- The heart of the scorpion is marked by **Antares**, a bright red **supergiant**.

- Scorpius contains rich **Milky Way** star fields, clusters, **nebulae**, and the strongest **X-ray** source in the sky, Scorpius X-1.

- In astrology, the dates for Scorpios are about 24 October–21 November (see **precession**).

season

Period of the year having a characteristic climate. The change in seasons is mainly due to the change in attitude of the Earth's axis in relation to the Sun, and hence the position of the Sun in the sky at a particular place.

- During winter, the Sun is low in the sky and has less heating effect because of the oblique angle of incidence and because the sunlight has further to travel through the **atmosphere**.

- The northern temperate **latitudes** have summer when the southern temperate latitudes have winter, and vice versa.

- In temperate latitudes four seasons are recognized: spring, summer, autumn and winter.

- Tropical regions have two seasons – the wet and the dry.

- Monsoon areas around the Indian Ocean have three seasons: the cold, the hot, and the rainy.

SETI

A programme originally launched by **NASA** in 1992, using powerful **radio telescopes** to search the skies for extraterrestrial signals. The name is an acronym for 'search for extraterrestrial intelligence'. Although NASA cancelled the SETI project in 1993, other privately funded SETI projects continue.

DIY SETI

Anyone with access to the Internet can participate in the seti@home project. After downloading the necessary software, the user's PC will 'number-crunch' blocks of data from the **Arecibo** radio telescope. The PC does the analysis in the background, while not busy with other tasks. Such a home computer user could get the credit for detecting the first signal from another civilization.

Seyfert galaxy

A **galaxy** with a small, bright centre, consisting of extremely hot gas moving at high speed around a massive central object, almost certainly a **black hole**. Almost all Seyferts are **spiral galaxies**. They seem to be closely related to **quasars**, but are about 100 times fainter. They are named after their discoverer US astronomer and astrophysicist Carl Seyfert (1911–1960).

singularity

A point in **space–time** at which the known laws of physics break down. A singularity is predicted to exist at the centre of a **black hole**, where gravitational forces become too intense to be described by existing theory. According to the **Big Bang** model of the origin of the universe, the expansion of the universe began from a singularity – which is to say little more than that the physics of today cannot deal with it. The English mathematician Roger Penrose (1931–) has shown that singularities arise in many circumstances in **relativity** theory, but argues that they are acceptable since they are always 'decently concealed' within black holes, so that they are not observable from the outside universe.

Sirius or the Dog Star or Alpha Canis Majoris

The brightest fixed star in the sky, lying in the constellation **Canis Major**.

- Sirius is 8.6 **light years** from the Sun.
- It is a hot, white star with a mass 2.3 times that of the Sun, a diameter 1.8 times that of the Sun, and a true **luminosity** 23 times that of the Sun.

- The name 'Sirius' is derived from the Greek word *seirios*, 'scorching'.
- In ancient Egypt, where its hieroglyph was a dog, its reappearance in the early morning sky heralded the annual rising of the Nile.

THE PUP

Sirius is orbited every 50 years by a **white dwarf**, Sirius B, also known as the Pup. This eighth-**magnitude** companion was first detected by German astronomer Friedrich Bessel (1784–1846) from its gravitational effect on the **proper motion** of Sirius. It was seen for the first time in 1862 but it was only in the 1920s that it was recognized as the first known example of a white dwarf.

Skylab
US **space station**, launched on 14 May 1973, which stayed in orbit for six years.

- *Skylab* was made from the adapted upper stage of a *Saturn V* rocket.
- At 75 tonnes/82.5 tons, it was the heaviest object ever put into space, and was 25.6 m/84 ft long.
- *Skylab* contained a workshop for carrying out experiments in weightlessness, an **observatory** for monitoring the Sun, and cameras for photographing the Earth's surface.
- Damaged during launch, it had to be repaired by the first crew of astronauts to board it.
- Three crews, each of three astronauts, occupied *Skylab* for periods of up to 84 days, at that time a record duration for human spaceflight.
- *Skylab* finally fell to Earth on 11 July 1979, dropping debris on Western Australia.

SOHO (Solar and Heliospheric Observatory)
A space probe launched in 1995 by the **European Space Agency** to study the **solar wind** of atomic particles streaming towards the Earth from the Sun. It also observes the Sun in ultraviolet and visible light. SOHO is operated jointly with **NASA** and cost $1.2 billion.

- *SOHO* is positioned at a stable **Lagrange point** 1.5 million km/938,000 mi from Earth towards the Sun.

- *SOHO* carries equipment for 11 separate experiments, including the study of the Sun's **corona**, measurement of its **magnetic field**, and observation of the **solar wind**.

- The Coronal Diagnostic Spectrometer (CDS) detects radiation at extreme ultraviolet wavelengths.

- The Michelson Doppler Imager (MDI) measures the **Doppler effect** in light wavelengths and can detect slight **solar oscillations**, vibrations on the Sun's surface that can reveal details of the structure of the Sun's interior.

- The Extreme-Ultraviolet Imaging Telescope (EIT) investigates the mechanisms that heat the Sun's corona.

- The Large-Angle Spectroscopic Coronagraph (LASCO) images the corona by detecting sunlight scattered by the coronal gases.

DISASTER AVERTED

SOHO's hydrazine fuel froze in June 1998, causing contact with it to be lost. Ground control at Goddard Space Flight Center, Greenbelt, Maryland USA, finally regained command of *SOHO* in September 1998, when the probe was turned so that its solar power arrays faced the Sun. Some data and instruments may have been permanently damaged by exposure to extreme temperatures.

---

solar cell
A semiconductor device that generates electrical power from the energy of sunlight. Solar cells are important in space research as a means of powering spacecraft. Large arrays of solar cells form 'wings' that extend from most artificial **satellites** and have also powered scientific experiments on the Moon and **Mars**.

solar eclipse
A brief darkening of the Sun that takes place when the Moon passes in front of it. It can happen only at new Moon, when the illuminated side of the Moon is turned completely away from the Earth (see **phase**).

- During a **total eclipse** the Sun's **corona** can be seen.

- A total solar eclipse can last up to 7.5 minutes.

- When the Moon is at its farthest from Earth it does not completely cover the face of the Sun, leaving a ring of sunlight visible. This is an **annular eclipse** (from the Latin word *annulus* meaning 'ring').

- Between two and five solar eclipses occur each year but each is visible only from a specific area.

- A total solar eclipse visible from south-western England took place on 11 August 1999 and lasted for two minutes. This was the first total solar eclipse to be visible from the UK since 1927; the next will be in 2090.

solar oscillations

Rapid vibrations of the Sun, or 'sunquakes', caused by sound waves travelling back and forth through the interior. They are detected by the **Doppler effect** on light from the Sun – the movement of the Sun's surface causes alterations in the wavelengths of the light, detectable by **spectroscopy**.

Solar System

The Sun (a star) and all the bodies orbiting it: the nine **planets** (**Mercury**, **Venus**, Earth, **Mars**, **Jupiter**, **Saturn**, **Uranus**, **Neptune**, and **Pluto**), their moons, the **asteroids**, and the **comets**.

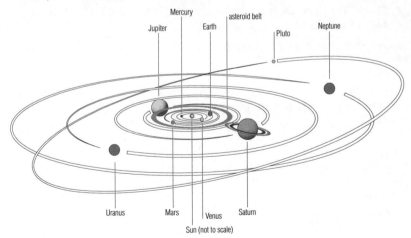

Solar System *Most of the objects in the Solar System lie close to the plane of the ecliptic. The planets are tiny compared to the Sun. If the Sun were the size of a basketball, the planet closest to the Sun, Mercury, would be the size of a mustard seed 15 m/48 ft from the Sun. The most distant planet, Pluto, would be a pinhead 1.6 km/ 1 mi away from the Sun. The Earth, which is the third planet out from the Sun, would be the size of a pea 32 m/100 ft from the Sun.*

- The Sun contains 99.86% of the mass of the Solar System.

- The objects in the Solar System have a common origin and development, as shown by these facts: the system is isolated in space; all the planets go round the Sun in **orbits** that are nearly circular and coplanar, and in the same direction in which the Sun itself rotates; and this same pattern is continued, with a few exceptions, in the moons of **Jupiter**, **Saturn**, **Uranus**, and **Neptune**.

- The Solar System formed by condensation from a cloud of gas and **dust** in space about 4.6 billion years ago.

solar time
The time of day as determined by the position of the Sun in the sky.

- *Apparent solar time*, the time given by a **sundial**, is not uniform because of the varying speed of the Earth in its elliptical **orbit**.

- *Mean solar time* is a uniform time that coincides with apparent solar time at four instants through the year. The difference between them is known as the **equation of time**, and is greatest in early November when the Sun is more than 16 minutes fast on mean solar time.

- Mean solar time on the Greenwich **meridian** is known as **Greenwich Mean Time** and is the basis of civil timekeeping.

  See also: *time measurement.*

solar wind
A stream of atomic particles, mostly **protons** and **electrons**, from the Sun's **corona**, flowing outwards at speeds of between 300 kps/200 mps and 1,000 kps/600 mps. The fastest streams come from 'holes' in the Sun's corona that lie over areas where no surface activity occurs. The solar wind pushes the gas of **comets'** tails away from the Sun, and 'gusts' in the solar wind cause geomagnetic disturbances and **aurorae** on Earth.

solstice
1 Either of the days on which the Sun is farthest north or south of the celestial equator each year. The summer solstice, when the Sun is farthest north, occurs around 21 June; the winter solstice around 22 December.

2 The positions on the **celestial sphere** at which the Sun is farthest north or south.

  See also: *equinox.*

Soyuz

A Soviet series of spacecraft, capable of carrying up to three cosmonauts. *Soyuz* (the name is Russian for 'union') spacecraft consist of three parts: a rear section containing engines; the central crew compartment; and a forward compartment that gives additional room for working and living space. They are now used for ferrying crews up to **space stations**, though they were originally used for independent space flight.

- *Soyuz 1* crashed on its first flight in April 1967, killing the lone pilot, Vladimir Komarov.

- In 1971 the three-man crew of *Soyuz 11* died on re-entry.

- In 1975 the **Apollo**–*Soyuz* test project resulted in a successful docking between the two spacecraft in **orbit**.

spaceflight

Travel outside the Earth's **atmosphere**. Serious discussion of the science and technology involved began with Russian scientist Konstantin Tsiolkovsky (1857–1935), who published a paper in 1903 on the use of **rockets** for travel into space. His work was purely theoretical; in the 1920s and 1930s pioneers, mainly in Germany and the United States, experimented with real liquid-fuelled rockets. The German V-2 ballistic missile of World War II was one outcome of this work, and was in turn the basis for rockets built by the United States and the Soviet Union after the war in their frantic 'space race'. Spaceflight turned from dream to reality on 4 October 1957, when the Soviet *Sputnik 1*, the first artificial **satellite**, orbited the Earth.

> ❝ For a third of the cost of a Hollywood movie, you can explore interplanetary space. ❞
>
> **Scott Hubbard**, Manager of the NASA Lunar Prospector mission

space industry

Commerce and manufacturing that take place wholly or partly outside the **atmosphere**.

- A major industry is already space-based: telecommunications via **communications satellite**, which represents a major part of global data traffic.

- Images from Earth **resources satellites** and even from **spy satellites** are now sold commercially to companies and governments.

- Manufacturing in space is still at the experimental stage. In **free fall**, or weightlessness, crystals can be grown that are almost completely free of the defects caused by convection currents on the Earth. This means that extremely high-quality medical drugs and semiconductors for electronics could be made in **orbit**.

- Sightseeing trips into orbit, space hotels and holidays on the Moon are real possibilities for the 21st century.

- Although extracting minerals from the Moon or from **asteroids** may need to be carried out as part of the exploration of the Moon and beyond, they may never be profitable industries, given the huge cost of transporting the product to Earth.

Spacelab

A small **space station** built by the **European Space Agency**, carried in the cargo bay of the US **space shuttle.**

- *Spacelab* remains in the shuttle cargo bay throughout each flight, returning to Earth with the shuttle.

- The laboratory consists of a pressurized module in which astronauts can work, and a series of pallets, open to the vacuum of space, on which equipment is mounted.

- *Spacelab* is used for **astronomy**, Earth observation, and experiments involving weightlessness and high vacuum.

- The pressurized module can be flown with or without pallets, or the pallets can be used on their own, in which case the astronauts remain in the shuttle's own crew compartment.

- All the sections of *Spacelab* can be reused many times.

- The first *Spacelab* mission, consisting of a pressurized module and pallets, lasted ten days during November and December 1983.

space shuttle

Any reusable crewed spacecraft. The first space shuttle was launched on 12 April 1981 by the USA. It was developed by **NASA** to reduce the cost of using space for commercial, scientific, and military purposes. After leaving its payload in space, the space-shuttle orbiter can be flown back to Earth to land on a runway, and is then available for reuse.

- NASA built four orbiters originally: *Columbia, Challenger, Discovery,* and *Atlantis.*

space shuttle *The launch of the space shuttle Challenger in 1985.*

- *Challenger* was destroyed in a mid-air explosion just over a minute after its tenth launch on 28 January 1986, killing all seven crew members onboard. It was the result of a failure in one of the solid **rocket** boosters. Flights resumed with redesigned boosters in September 1988.

- A replacement orbiter, *Endeavour*, was built, which had its maiden flight in May 1992.

THE ORBITER

The space-shuttle orbiter, the part that goes into space, is 37.2 m/122 ft long and weighs 68 tonnes. Two to eight crew members occupy the orbiter's nose section, and missions last up to 30 days. In its cargo bay the orbiter can carry up to 29 tonnes of satellites, scientific equipment, *Spacelab*, or military payloads. At launch, the shuttle's three main engines are fed with liquid fuel from a cylindrical tank attached to the orbiter; this tank is discarded shortly before the shuttle reaches orbit. Two additional solid-fuel boosters provide the main thrust for launch, but are jettisoned after two minutes.

space station

Any large structure designed for human occupation for extended periods of time in orbit around the Earth, or, in the future, other worlds. Space stations are used for carrying out astronomical observations and surveys of Earth, as well as for biological studies and the processing of materials in weightlessness. The first space station was **Salyut 1**. The United States later launched **Skylab**. The Soviet **Mir** was launched in 1986 and operational until 1999. The **International Space Station** is now under construction.

spacesuit

The protective suit worn by astronauts and cosmonauts in space. It provides an insulated, air-conditioned cocoon in which the wearer can live and work for hours at a time while outside the spacecraft.

- The suit provides air to breathe, and removes exhaled carbon dioxide and moisture.

- The suit's outer layers insulate the occupant from the extremes of hot and cold in space (–150°C/–240°F in the shade to +180°C/+350°F in sunlight).

- Inside the suit is a cooling garment that helps to keep the body at a comfortable temperature even during vigorous work.

- The suit also provides protection from the impact of small **meteoroids**.

space-time

The combination of space and time, which the **relativity** theory of Albert **Einstein** revealed to be intimately interconnected. Space and time are both altered by a system's state of motion and by whether it is in a gravitational field or not. The ways in which this happens are best described by the mathematics of a four-dimensional geometry. Space-time is viewed in relativity as actively influencing the behaviour of bodies moving through it, and in turn being influenced by the matter in it; for example, the space-time around the Sun is distorted by the presence of the Sun, with the result that the paths of rays of **light** are bent.

spectroscopy

The analysis of **light** and other electromagnetic radiation to determine what proportion of each wavelength is present. Originally light was split up using **prisms**; later the **diffraction** grating, a metal or glass plate engraved with closely spaced lines, was used instead. Similar devices are used for **infrared radiation** and **ultraviolet radiation**, while radio waves, **X-rays** and **gamma-rays** are analysed electronically. Dispersing radiation into its different wavelengths produces a band called a **spectrum**.

- Instruments used to analyse light are called *spectrometers, spectroscopes*, or *spectrographs*.

- Particular wavelengths are often called *lines* in spectroscopy, because they appear as bright or dark lines across a spectrum – the line is an image of the slit through which light enters the spectrometer.

- *Emission lines* are bright lines in a spectrum, showing wavelengths being given out by the light source. In solids and liquids these lines are often 'smeared' into bands. Gases usually give out sharper lines, which makes it possible to identify the elements and compounds that the gas consists of.

- ***Absorption lines*** are dark lines crossing a spectrum. They show where cooler gases are absorbing light of those wavelengths from the light of a hotter source beyond. Thus the absorption lines in a star's spectrum show where light coming from the hot surface of the star is being absorbed by the star's cooler **atmosphere**, or possibly by **interstellar matter**.

- The lines in a **spectrum** can be shifted because of the motion of the emitting or absorbing material. This **Doppler effect** causes the wavelengths to lengthen when the motion is away from us, and to shorten when it is towards us.

  See also: *red-shift.*

Spectroscopic binaries
A spectroscopic binary is a **binary system** in which the two stars are so close together that they cannot be seen separately, but their spectra can be distinguished by a spectroscope. As the two stars revolve around their mutual centre of mass, they alternately approach and recede from the observer, resulting in a periodic Doppler shift (see **Doppler effect**) in the lines of their spectra. If the orbital motion happens to lie at right angles to the line of sight, there is no approach nor recession so that such a system cannot be detected as a spectroscopic binary.

- In most cases, only the spectrum of one star can be distinguished, so one set of spectral lines is seen, moving back and forth periodically.

- In about one case in six, the component stars are sufficiently similar in brightness for the spectra of both to appear; then as one star approaches, the other recedes, so that the spectral lines split and rejoin.

- The line-of-sight velocity of one star or both can be calculated from the movement of the lines.

- Analysis of these velocities gives information about the masses of the components of the binary.

- The first spectroscopic binary to be discovered was the brighter component of Mizar in the **constellation Ursa Major** by E C Pickering (1846–1919) in 1889.

- Many hundreds are now known, with orbital periods ranging from 82 minutes to 15 years. The most common **period** is a few days.

Arranging the stars

Stars are classified according to their surface temperature and **luminosity**, as determined from their spectra. They are assigned a spectral type (or class) denoted by the letters O, B, A, F, G, K, and M, where O stars (about 40,000°C/72,000°F) are the hottest and M stars (about 3,000°C/5,400°F) are the coolest.

Each type may be further divided into ten subtypes by adding a numeral to the type letter: B0, B1, B2, and so on. Stars are also assigned a luminosity class, denoted by a Roman numeral attached to the spectral type: I (**supergiants**), II (bright giants), III (giants), IV (subgiants), V (main sequence), VI (subdwarfs), or VII (**white dwarfs**). The Sun is classified as type G2V. Stars are assigned to positions in the **Hertzsprung–Russell diagram** on the basis of this classification.

SPECTRAL CLASSIFICATION

To remember the spectral classification sequence: the initial letters of

Octopus brains, a favourite gastronomic kitchen menu, require no sauce.

or

On, backward astronomer, Forget geocentricity!: Kepler's motions reveal nature's simplicity.

are the letters in the sequence:

O, B, A, F, G, K, M, (R, N, S,)

ranging from hotter stars, O, to cooler ones, M. (The classes R, N, and S are generally no longer used, but appear in most of the mnemonics.)

---

spectrum

The pattern obtained when waves – of electromagnetic radiation, sound or any other kind – are split into their different wavelengths. An example of a spectrum of sunlight is the coloured band obtained when the **light** is passed through a **prism**. More sophisticated instruments are used to study spectra in **spectroscopy**.

spherical aberration

A defect that impairs the image in an optical instrument. The image is blurred because different parts of a spherical lens or **mirror** have different focal lengths. Elaborate computer programs are now used to design lenses in which spherical aberration is minimized.

See also: *chromatic aberration.*

Spica or Alpha Virginis

The brightest star in the **constellation Virgo** and the 16th-brightest star in the night sky.

- Spica has a true **luminosity** over 1,500 times that of the Sun.

- It is a spectroscopic **binary star**, the components of which **orbit** each other every four days.

- It is 140 **light years** from the Sun.

- *Spica* is Latin for 'ear of corn'.

spiral galaxy

One of the main classes of **galaxy**, comprising up to 30% of known galaxies. Spiral galaxies are characterized by a central bulge surrounded by a flattened disc. The disc contains (normally) two spiral arms, composed of hot young stars and clouds of **dust** and gas. About half of spiral galaxies are **barred spirals**, in which the arms originate at the ends of a bar across the central bulge. The bar is not a rigid object but consists of stars in motion about the centre of the galaxy.

See also: *elliptical galaxy, irregular galaxy.*

Sputnik

- A series of ten Soviet Earth-orbiting **satellites**, taking their name from the Russian word meaning 'fellow traveller'. *Sputnik 1* was the world's first artificial **satellite**, launched on 4 October 1957. Later *Sputniks* were test flights of the *Vostok* spacecraft. *Sputniks* were superseded in the early 1960s by the *Cosmos* series of satellites.

- *Sputnik 1* weighed 84 kg/185 lb, with a 58-cm/23-in diameter.

- It carried only a simple radio transmitter, which allowed scientists to track it as it orbited the Earth.

- It burned up in the **atmosphere** 92 days later.

- *Sputnik 2*, launched on 3 November 1957, weighed about 500 kg/1,100 lb, including its payload, the dog Laika, the first living creature in space. There was no way to return her to Earth, and she died in space.

spy satellite

A **satellite** that can survey the Earth's surface in great detail to determine anything of military value – the location and nature of military equipment, terrain, industrial plant, and so on. They are usually in tilted orbits so that they travel large distances north and south of the equator to survey terrain of interest in foreign countries. They are at low altitudes, to see the greatest possible detail. Satellites with very powerful cameras can make out individual people on the ground and read markings on parked aircraft.

star

A luminous globe of gas, mainly **hydrogen** and **helium**, which produces heat and light by nuclear reactions. Stars have lifetimes ranging from tens of millions to tens of billions of years, and change greatly in appearance at different stages in their lives.

- The smallest mass possible for a star is about 8% that of the Sun (80 times that of **Jupiter**), otherwise nuclear reactions do not occur. Objects with less than this critical mass shine only dimly, and are termed **brown dwarfs**.

- Stars in the prime of life burn hydrogen and convert it into helium. At this stage they are called **main sequence** stars. The star remains practically unaltered while its hydrogen lasts.

- The star's position on the main sequence (see **Hertzsprung–Russell diagram**) and how long it remains there are determined by the mass of the star: the larger the mass, the brighter the star and the shorter its life.

- A star with the mass of the Sun remains on the main sequence for about 10 billion years, twice the present age of the Sun.

- The Sun is thus expected to remain at this stage for another 5 billion years.

- Surface temperatures of main-sequence stars range from 2,000°C/3,600°F to above 30,000°C/54,000°F.

- The corresponding colours range from red to blue-white.

- The nuclear reactions take place near the centre, so the star gradually acquires an inert helium core, surrounded by a thin shell of burning hydrogen.

- When all the hydrogen at the core of a main-sequence star has been converted into helium, the star swells to become a **red giant**, about 100 times its previous diameter and with a cooler, redder surface.

BIRTH OF A STAR

Stars are born when **nebulae** (giant clouds of **dust** and gas) contract under the influence of **gravity**. These clouds consist mainly of hydrogen and helium, with traces of other elements and dust grains. A huge volume of **interstellar matter** gradually begins to collapse, and the temperature and pressure in its core rises as the star grows smaller and denser. As the star is forming, it is surrounded by **EGGs** (evaporating gaseous globules), the oldest of which was photographed in the Eta Carinae nebula in 1996 by the **Hubble Space Telescope**.

At first the temperature of the **protostar** scarcely rises, as dust grains radiate away much of the heat, but as it grows denser less of the heat generated can escape, and it gradually warms up. At about 10 million °C/18 million °F the temperature is hot enough for a nuclear reaction to begin, and hydrogen nuclei fuse to form helium nuclei; vast amounts of energy are released, contraction stops, and the star begins to shine. A star with a mass roughly equal to that of our Sun takes a few million years to reach the main sequence (begin hydrogen-burning).

Death of a star

How a star ends its life depends on its mass. If its mass is less than 1.2 that of the Sun:

- When it leaves the main sequence and becomes a giant star, its outer layers drift off into space to form a **planetary nebula**.

- Its core collapses in on itself to form a small and very dense body called a **white dwarf**.

- Eventually the white dwarf fades away, leaving a nonluminous dark body.

If the mass is greater than 1.2 that of the Sun:

- The star does not end as a white dwarf, but passes through its life cycle quickly, becoming a red **supergiant**.

- As the star's core grows hotter, further nuclear transformations take place, resulting in the

THE BRIGHTEST STAR

The brightest known star is the Pistol Star, discovered near the centre of the **Milky Way** in 1997 by the **Hubble Space Telescope**. It is 10 million times brighter than the **Sun** and 100 times larger, and emits as much energy in seconds as the Sun does in one year.

helium being converted first into carbon and oxygen, and then into heavier elements, and finally iron.

- The star eventually explodes into a brilliant **supernova**.

- Part of the core remaining after the explosion may collapse to form a small superdense star, consisting almost entirely of **neutrons** and therefore called a **neutron** star.

- Neutron stars, also called **pulsars**, spin very quickly, giving off pulses of radio waves.

- If the collapsing core of the supernova has a mass more than twice that of the Sun it does not form a neutron star; instead it forms a **black hole**, a region so dense that neither matter nor radiation can escape from it.

SAKURAI'S OBJECT

A new star, named Sakurai's Object (after the amateur Japanese astronomer who discovered it in 1996 in the **constellation** of **Sagittarius**), has been undergoing an extraordinarily rapid evolution since its discovery. It expanded from being an Earth-sized hot dwarf, with a surface temperature of 50,000 °C/90,032°F, to being a bright yellow supergiant about 80 times wider than the Sun and no hotter than 6,000 °C. Sakurai's Object may be a red giant star that had previously shrunk, the contraction of its core triggering nuclear reactions and subsequent reinflation.

See also: *binary star, variable star.*

steady-state theory
A theory that formerly rivalled the **Big Bang** theory. It claimed that the universe has no origin but is eternally unchanging on the large scale. As it expands new matter is created continuously throughout space. The theory was proposed in 1948 by Austrian-born British cosmologist Hermann Bondi (1919–   ), Austrian-born US astronomer Thomas Gold (1920–   ), and Fred **Hoyle**. it was a stimulus to the development of theories of the synthesis of elements within stars, and was thus a corrective to the then prevalent belief that all the elements were formed in the Big Bang. However, it was dealt a severe blow in 1964 by the discovery of **cosmic background radiation** (radiation left over from the formation of the universe) and in its original form is now rejected.

stony-iron meteorite
One of the main categories of **meteorites**, the fragments remaining from the fall of large **meteoroids**. They consist of a mixture of roughly equal quanti-

ties of rock and iron. They probably originate at the boundary between the metallic core and rocky crust of some large **asteroid** that was broken up by collisions. They represent only about 1% of observed meteorite falls.

stratosphere

A layer of the Earth's **atmosphere** which lies above the **troposphere**, the lowest layer, and in which temperature increases with altitude.

- The stratosphere is from 10 km/6 mi to 50 km/31 mi high.

- Its temperature increases from –60°C/–140°F to near 0°C/32°F.

- The top of the stratosphere, where there is a temperature maximum, is called the stratopause. Here, photons of **ultraviolet radiation** are absorbed by molecules to form new gases, especially **ozone** from oxygen.

- Ozone is a better absorber of ultraviolet radiation than is ordinary oxygen, and it is the **ozone layer** within the stratosphere that prevents lethal amounts of ultraviolet radiation from reaching the Earth's surface.

Sun

The star at the centre of the **Solar System**. It is composed of about 70% **hydrogen** and 30% **helium**, with other elements making up less than 1%. The Sun's energy is generated by nuclear fusion reactions that turn **hydrogen** into **helium** at its centre. The core is far denser than mercury or lead on Earth.

The Sun is about 4.7 billion years old, with a predicted total lifetime of 10 billion years. At the end of its life, it will expand to become a **red giant** the size of the **orbit** of **Mars**, then shrink to become a **white dwarf**.

THE TURBULENT SUN

Sometimes bright eruptions called **flares** occur near sunspots. Above the bright visible surface, called the **photosphere**, lies a layer of thinner gas called the **chromosphere**, visible only by means of special instruments or at **eclipses**. Tongues of gas called **prominences** extend from the chromosphere into the **corona**, a halo of hot, tenuous gas surrounding the Sun. Gas boiling from the corona streams outwards through the Solar System, forming the **solar wind**. Activity on the Sun, including sunspots, flares, and prominences, waxes and wanes during the solar cycle, which peaks every 11 years or so, and is connected with the solar **magnetic field**. The further out from the Sun, the hotter it is: the Sun's corona is 1,994,000 °C/3,589,000 °F hotter

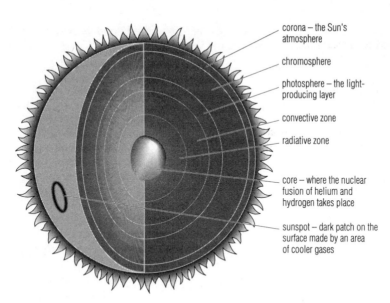

corona – the Sun's atmosphere

chromosphere

photosphere – the light-producing layer

convective zone

radiative zone

core – where the nuclear fusion of helium and hydrogen takes place

sunspot – dark patch on the surface made by an area of cooler gases

Sun *The structure of the Sun. Nuclear reactions at the core release vast amounts of energy in the form of light and heat that radiate out to the photosphere and corona. Surges of glowing gas rise as prominences from the surface of the Sun and cooler areas, known as sunspots appear as dark patches on the giant star's surface.*

The Sun spins on its axis every 25 days near its equator, but more slowly towards its poles. Its rotation can be followed by watching the passage of dark **sunspots** across its disc.

The wall
A wall of heated **hydrogen** atoms at temperatures of 20,000–40,000°C/ 35,500–71,500°F) that forms in the path of the Sun as it moves through space was discovered in 1995. The wall lies about 2,240 million km/1,555 million mi from the Sun, and its existence had been predicted by theorists.

WATER IN A HOT PLACE

US astronomers reported finding water on the Sun in 1995. The water, in the form of superheated steam, was located in two sunspots where the temperature was 'only' 3,000°C/5,400°F) (as opposed to 5,500°C/9,932°F elsewhere on the surface).

SUN: STATISTICS

Diameter	Mass	Mass (Earth = 1)	Energy output
1,392,000 km/ 865,000 mi	1.99 x 1027 tonnes	328,910	3.83 x 1023 kW

Rotation period (equatorial)	Temperature (surface)	Temperature (centre) (water = 1)	Average density	Volume (Earth = 1)
25 days (equator)	5,500 °C/ 9,900 °F	15,000,000 °C/ 27,000,000 °F	1.41	1,304,000

sundial

Instrument that measures time by means of a shadow cast by the Sun. Almost completely superseded by the proliferation of clocks, it survives ornamentally in gardens. The dial is marked with the hours at graduated distances, and a style or gnomon (parallel to Earth's axis and pointing to the north) casts the shadow.

See also: *time measurement.*

sunspot

An area on the surface of the Sun that is dark by comparison with its surroundings. It is cooled by strong **magnetic fields** that block the outward flow of heat to the Sun's surface.

- Sunspots consist of a dark central **umbra**, at about 3,700°C/6,700°F, and a lighter surrounding **penumbra**, at about 5,200°C/9,400°F.

- They last from several days to over a month, ranging in size from 2,000 km/1,250 mi to groups stretching for over 100,000 km/62,000 mi.

- Sunspots are more common during active periods in the Sun's magnetic cycle, when they are sometimes accompanied by **flares**.

- The number of sunspots visible at a given time varies from none to over 100, in a cycle averaging 11 years.

- There was a lull in sunspot activity, known as the Maunder minimum, from 1645 to 1715, that coincided with a cold spell in Europe.

supercluster

A grouping of several **galaxy clusters** forming a structure about 100–300 million **light years** across. Our own **Galaxy** and its neighbours lie on the edge of the **Local Supercluster**, of which the **Virgo** cluster is the dominant member.

supergiant
The largest and most luminous type of star known, with a diameter up to 1,000 times that of the Sun and absolute **magnitude** between –5 and –9. Supergiants are likely to become **supernovae**.

superior planet
A **planet** that is further away from the Sun than the Earth is: that is, any one of **Mars**, **Jupiter**, **Saturn**, **Uranus**, **Neptune**, and **Pluto**.
   See also: *inferior planet.*

supernova
The explosive death of a star, which temporarily attains a brightness of 100 million Suns or more and shines as brilliantly as a small **galaxy** for a few days or weeks. Very approximately, it is thought that a supernova explodes in a large galaxy about once every 100 years. Many supernovae – astronomers estimate some 50% – remain undetected because they are dimmed by interstellar **dust**.

- Gas ejected by such an explosion forms an expanding radio source.

- Supernovae are the source of elements heavier than **hydrogen** and **helium**, which are scattered through the **interstellar material** from which new **planetary systems** form.

- A supernova at maximum attains an absolute **magnitude** of –14 to –20, over 10,000 times brighter than an ordinary **nova**. When one occurs in a galaxy its **light** quite often equals or exceeds the combined light of all the other stars and **nebulae** in the system, while its **spectrum** indicates very violent internal motions.

TYPES OF SUPERNOVA

The name 'supernova' was coined by the US astronomers Fritz Zwicky (1898–1974) and Walter Baade (1893–1960) in 1934. Zwicky was also responsible for the division into types I and II:

- Type I supernovae are thought to occur in **binary star** systems, in which gas from one member of the system falls on to the other, a **white dwarf**, causing it to explode.

- Type II supernovae occur in stars ten or more times as massive as the Sun, which suffer runaway internal nuclear reactions at the ends of their lives, leading to explosions. These are thought to leave behind **neutron stars** and **black holes**.

Surveyor missions

A series of seven US robot Moon landers launched in the period 1966–68 that paved the way for the *Apollo* missions. Each craft was equipped with TV cameras and a scoop for examining lunar soil. With the exceptions of *Surveyor 2* and *Surveyor 4*, all the landings were successful.

- *Surveyor 1* landed in the Ocean of Storms in June 1966, becoming the first US spacecraft to make a soft landing on the Moon.

- *Surveyor 3*, which landed in the Ocean of Storms in April 1967, was visited by the crew of *Apollo 12* in November 1969.

synodic period

The time taken for a **planet**, the Moon or some other celestial body to return to the same position in its orbit as seen from the Earth – for example, from one **conjunction** to the next. It differs from the **sidereal period** because the Earth is moving in **orbit** around the Sun.

See also: *phase.*

Taurus

A conspicuous **zodiacal constellation** in the northern hemisphere.

- The V-shaped Hyades **open cluster** forms the bull's head, with **Aldebaran** as the red eye.

- The Pleiades open cluster is in the shoulder.

- Taurus also contains the **Crab Nebula**, the remnants of the **supernova** of AD 1054, which is a strong radio and **X-ray** source and the location of one of the first **pulsars** to be discovered.

- T Tauri is one of a class of **variable** stars thought to represent an early stage of stellar formation.

- The Sun passes through Taurus from mid-May to late June.

- In astrology, the dates for Taurus are between about 20 April and 20 May (see **precession**).

telescope

An optical instrument that magnifies images of faint and distant objects; any device for collecting and focusing **light** and other forms of electromagnetic radiation. It is the major research tool of **astronomy**.

The refractor

In a refractor, light is collected by the objective lens, which focuses light down a tube, forming an image magnified by an eyepiece. Invention of the refractor is attributed to a Dutch optician, Hans Lippershey (*c.* 1570–*c.* 1619), in 1608. Hearing of the invention in 1609, the Italian physicist **Galileo**

> ### THE IMPORTANCE OF APERTURE
>
> Astronomical telescopes are always described in terms of their aperture, which is the breadth of the objective – the main lens or main **mirror**. A telescope with a large aperture can distinguish finer detail than one with a small aperture. It can also detect fainter objects, because it has more light-gathering power. Magnification depends on the length of the light-path from the objective to the image.

quickly constructed one for himself and went on to produce a succession of such instruments, which he used from 1610 onwards for epoch-making astronomical observations. The largest refracting telescope in the world, at Yerkes Observatory, Wisconsin, USA, has an aperture of 102 cm/40 in.

The reflector
In a reflector, light is collected and focused by a concave mirror. The first reflector was built about 1670 by the English physicist Isaac **Newton**. Large mirrors are cheaper to make, easier to mount, and easier to correct for aberrations than large lenses, so all the largest telescopes are reflectors. The largest reflector with a single mirror, 6 m/236 in, is at Zelenchukskaya, Russia. Telescopes with larger apertures composed of numerous smaller segments have been built, such as the **Keck Telescopes** on Mauna Kea, Hawaii.

A hybrid telescope
The catadioptric telescope is a combination of lenses and mirrors. Schmidt telescopes are used for taking wide-field photographs of the sky. Invented in 1930 by an Estonian astronomer, Bernhard Schmidt (1879–1935), they have a main mirror plus a thin, carefully shaped lens at the front of the tube to increase the field of view without introducing **spherical aberration**. Exam- ples are the 1. 2-m/48-in Schmidt telescope on **Mount Palomar** and the UK Schmidt telescope, of the same size, at Siding Spring in Australia.

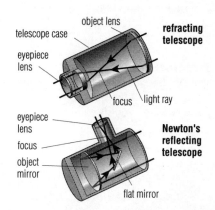

telescope *Refracting and reflecting telescopes. The refracting telescope uses a large objective lens to gather light and form an image which the smaller eyepiece lens magnifies. A reflecting telescope uses a mirror to gather light.*

Liquid mirrors
The liquid-mirror telescope is a reflecting telescope constructed with a rotating mercury mirror. In 1995 **NASA** completed a 3-m/9.8-ft liquid mirror telescope at its Orbital Debris Observatory in New Mexico, USA.

See also: *Cassegrain telescope, radio telescope, scintillation.*

TELESCOPES IN SPACE

Large telescopes placed in **orbit** are above the distorting effects of the Earth's **atmosphere**. Telescopes carried on **satellites** have been used to study **infrared radiation**, **ultraviolet radiation**, and **X-rays**, which do not penetrate the atmosphere but carry information about the births, lives, and deaths of stars and **galaxies**. The 2.4-m/94-in **Hubble Space Telescope**, launched in 1990, can see the sky more clearly than can any telescope on Earth. It will be dwarfed by the planned Next Generation Space Telescope, which will be placed in orbit beyond the Moon, so that it can observe infrared radiation without interference from the heat radiation of the Earth.

terminator

The dividing line between the sunlit and dark hemispheres of the Moon, another **satellite**, or a **planet**. On the Moon, its irregular edge reveals the roughness of the terrain.

terraforming

The hypothetical process of transforming a **planet** – particularly **Mars** – into a form more resembling the Earth and more suitable for life. Such an alteration of Mars would require:

- Warming the planet by pumping greenhouse gases such as CFCs into the atmosphere; this would melt the **polar caps** and release water vapour and carbon dioxide into the **atmosphere**.

- Using fungi and lichens or other vegetation that would flourish on the carbon dioxide to produce oxygen, which would also create a protective **ozone layer**.

- Introducing nitrogen-fixing plants able to create a nourishing soil.

The whole process could take 100,000 years – and many would think an entire planet should not be manipulated in this way.

thermosphere

The highest layer of the Earth's **atmosphere**, 80 km/50 mi to about 700 km/ 435 mi high, in which temperature rises with altitude to extreme values of thousands of degrees. But high thermosphere temperatures represent little heat because they are defined by motions among very few and widely spaced atoms and molecules.

tides

The rise and fall of the oceans, and to a lesser extent of the solid land, because of the gravitational influence of the Moon and Sun. The effect arises because the strength of the gravitational field of each of these bodies is slightly different on the sides of the Earth towards and away from the attracting body. The oceans tend to pile up at those two places. High tide occurs, on average, at intervals of 12 hr 24 min 30 sec. The maximum high tides, or spring tides, occur at or near new and full Moon when the Moon and Sun are in line and exert the greatest combined gravitational

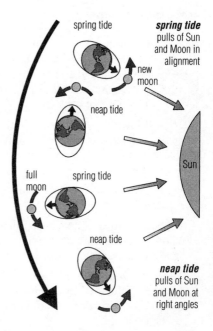

tides *The gravitational pull of the Moon is the main cause of the tides. Water on the side of the Earth nearest the Moon feels the Moon's pull and accumulates directly under the Moon. When the Sun and the Moon are in line, at new and full moon, the gravitational pull of Sun and Moon are in line and produce a high spring tide. When the Sun and Moon are at right angles, lower neap tides occur*

pull. Lower high tides, or neap tides, occur when the Moon is in its first or third quarter and the Moon and Sun are at right angles to each other (see **phase**). Tidal forces can be seen at work elsewhere in the universe – for example, in the distortion of the shapes of **galaxies** subjected to the gravitational pull of other galaxies

---

THE WARMING TIDE

Tidal heating is a process in which a celestial body is heated internally by tidal stresses set up by the gravitational pull of another body. Tidal heating occurs in some moons of the giant planets – notably, it is the heat source for the violent volcanic activity on **Io**, one of the moons of **Jupiter**.

time measurement

The measurement of time used to be based on the day, the time taken for one rotation of the Earth on its axis, but this was found to be irregular. Therefore the second, the standard scientific unit of time, was redefined in 1956 in terms of the year, the length of the Earth's annual **orbit** of the Sun. In 1967 it was again redefined in terms of the vibrations of a certain wavelength of the element caesium.

- The worldwide time standard, called UTC (Universal Coordinated Time) is determined by the International Bureau of Weights and Measures (BIPM), based in Paris, and this is transmitted to the world's national time authorities.

- From 1986 in scientific usage the term **'Greenwich Mean Time'** was replaced by 'UTC'.

- However, the Greenwich **meridian**, adopted as the **prime meridian** in 1884, remains that from which all **longitudes** are measured, and the world's standard time zones are calculated from it.

- National **observatories** make standard time available in their countries. In the UK, the BBC broadcasts six pips at certain hours (five short, from second 55 to second 59, and one long, the start of which indicates the precise minute).

- Atomic clocks have an accuracy greater than 1 second in 300 million years.

time travel

Travel into the future or past, much loved of science-fiction writers but also a scientific possibility in certain extreme situations in the universe.

A kind of travel into the future is well established in **relativity** theory: time passes more slowly in a fast-moving system, relative to stationary observers. For example, fast subatomic particles decay at a slower rate than slow ones. Interstellar explorers, travelling at large fractions of the speed of **light**, would return to Earth after, say, 10 years had passed according to their ship's clocks and their own bodily processes, to find that, say, 50 years had passed on Earth (depending on how fast they had moved). In this sense, they could be said to have 'travelled into the future'.

Space-time is distorted near massive spinning bodies, such as **pulsars**. In theory a spaceship passing near an extremely high-mass, extremely fast-spinning object could arrive in a different time on the other side. But no such object is yet known to astronomers.

'**Wormholes**' connecting **black holes** in different places and times are also much discussed. an object entering one mouth of the wormhole through the black hole could arrive in the past or future at the other end. But no way is known of preventing any traveller from being torn apart by the immense forces present, or of escaping intact from the black hole at the destination. Time travel is still just a hypothetical possibility for today's theoretical physicists.

See also: *parallel universes.*

Titan

The largest moon of the planet **Saturn** and the second-largest moon in the **Solar System** (**Ganymede**, a moon of **Jupiter**, is larger).

- Titan is the only moon in the Solar System with a substantial **atmosphere** (mostly nitrogen). This is topped with smoggy orange clouds that obscure the surface.

- The surface may be covered with liquid ethane lakes.

- The surface atmospheric pressure is greater than Earth's.

- Radar signals suggest that Titan has dry land as well as oceans.

- Titan was discovered in 1655 by the Dutch mathematician and astronomer **Christiaan Huygens**.

See also: *Cassini–Huygens mission.*

TITAN: STATISTICS		
Diameter	**Distance from centre of planet**	**Period**
5,150 km/3,200 mi	1,222,000 km/759,000 mi	15.95 days

Titius–Bode Law

An alternative name for **Bode's Law**.

total eclipse

A **solar eclipse** in which the Sun is completely hidden by the Moon, or a **lunar eclipse** in which the Moon completely enters the **umbra** of the Earth's shadow.

See also: *partial eclipse.*

Triton
The largest of **Neptune's** moons. It is slightly larger than the planet **Pluto**, which it is thought to resemble in composition and appearance.

- Triton revolves around Neptune in a **retrograde** (east to west) direction.

- It takes the same time to rotate about its own axis as it does to make one revolution of Neptune.

- Triton was probably formerly a separate body like Pluto but was captured by Neptune.

- Triton's surface, as revealed by the *Voyager 2* space probe, has a temperature of –235°C/–391°F, making it the coldest known place in the **Solar System**.

- It is covered with frozen nitrogen and methane, some of which evaporates to form a tenuous **atmosphere** with a pressure only 0.00001 that of the Earth at sea level.

- Triton has a pink south **polar cap**, probably coloured by the effects of solar radiation on methane ice.

- Dark streaks on Triton are thought to be formed by geysers of liquid nitrogen.

- The surface has few impact **craters** (the largest is the Mazomba, with a diameter of 27 km/17 mi), indicating that many of the earlier craters have been erased by the eruption and freezing of water (cryovulcanism, or 'cold vulcanism').

- Triton was discovered in 1846 by the British astronomer William Lassell (1799–1880) only weeks after the discovery of Neptune.

TRITON: STATISTICS

Diameter	Distance from centre of planet	Period
2,700 km/1,680 mi	355,000 km/221,000 mi	5.88 days

Trojan asteroids
A group of **asteroids** that share the orbit of **Jupiter**. Some, the **Achilles group**, move around a point 60° ahead of the **planet**; others, the **Patroclus group**, move around a point 60° behind the planet

tropics

1 On the **celestial sphere**, the circles of **declination** (circles parallel to the celestial equator) that mark the farthest distance north or south that the Sun can reach during the year. The tropic of Cancer is the northern tropic, the tropic of **Capricorn** is the southern.

2 On the Earth, the tropics are the parallels of **latitude** directly beneath the celestial tropics, and lie approximately 23° 30′ N and S of the equator. They are the limits of the area of the Earth's surface in which the Sun can be directly overhead at some time during the year.

See also: *equinox, solstice.*

troposphere

The lowest level of the Earth's **atmosphere** (altitude ranging up to 10 km/ 6 mi). It is heated to an average temperature of 15°C/59°F by the Earth, which in turn is warmed by infrared and visible radiation from the Sun. Warm air cools as it rises in the troposphere and this rising of warm air causes rain and most other weather phenomena. The top of the troposphere is at approximately –60°C/ –76°F.

21-centimetre line

An important wavelength in **radio astronomy**. Atoms of ordinary **hydrogen**, the most abundant type of atom in the universe, emit radio waves of 21 cm/ 9.84 in wavelength. This makes the structure of our own **Milky Way Galaxy** and other nearby **galaxies** 'visible' to **radio telescopes**.

ultraviolet astronomy

The study of cosmic **ultraviolet radiation** using artificial **satellites**. Only a tiny percentage of solar ultraviolet radiation penetrates the **atmosphere**, this being the less dangerous longer-wavelength ultraviolet. The dangerous shorter-wavelength radiation is absorbed by gases in the **ozone layer** high in the Earth's upper **atmosphere**.

ORBITING ASTRONOMICAL OBSERVATORY (OAO) SATELLITES

Various **satellites** have provided scientists with a great deal of information regarding cosmic ultraviolet emissions.

- There were several US Orbiting Astronomical Observatory (*OAO*) satellites. *OAO*-1, launched in 1966, failed after only three days.
- *OAO*-2, put into **orbit** in 1968, operated for four years instead of the intended one year, and carried out the first ultraviolet observations of a **supernova** and also of **Uranus**.
- *OAO*-3 (*Copernicus*), launched in 1972, continued transmissions into the 1980s and discovered many new ultraviolet sources.
- The International Ultraviolet Explorer (*IUE*), which was launched in January 1978 and ceased operation in September 1996, observed all the main objects in the **Solar System** (including Halley's comet), stars, **galaxies**, and **interstellar matter**.

---

ultraviolet radiation

Electromagnetic radiation invisible to the human eye, of wavelengths from about 400–4 nanometres, where the **X-ray** range begins. (One nanometre is one-billionth of a metre.) The radiation may be detected with ordinary photographic plates or films, but more sophisticated detectors are used in **astronomy**. Shorter-wavelength ultraviolet (UV) radiation is absorbed by the **ozone layer**, but levels of ultraviolet radiation rose an average of 6.8% per decade in the northern hemisphere and 9.9% in the southern hemisphere in the period 1972–96, according to data gathered by the Total Ozone Mapping Spectrometer on the *Nimbus* 7 satellite.

Ulysses

A space probe to study the Sun's poles, launched in 1990 by a US **space shuttle**. It is a joint project by **NASA** and the **European Space Agency**. In February 1992, the gravity of **Jupiter** swung *Ulysses* on to a path that looped it first under the Sun's south pole in 1994 and then over its north pole in 1995 to study the Sun and **solar wind** at **latitudes** not clearly observable from the Earth.

umbra

1 The inner part of a shadow, which receives no light directly from the source of illumination. When the Moon passes through the umbra of the Earth's shadow, there is a total **lunar eclipse**. In a **solar eclipse**, a **total eclipse** is seen from those places where the umbra of the Moon's shadow falls.

2 The inner and darker part of a **sunspot**.

universe, future of

The ultimate future of the universe depends on what happens to the expansion that is taking place today. A central problem of cosmology is whether the mutual recession of the **galaxies** is slowing, accelerating, or staying the same. The mutual gravitational attraction of the galaxies is tending to slow the galaxies' movement, but the observed quantities of matter in the universe are insufficient to stop it completely. However, it is certain that there is much more matter in the universe, in the form of **dark matter**, and if there were enough, it could eventually (after billions or even trillions of years) reverse the expansion. One theory holds that there will be a Big Crunch, like the **Big Bang** run in reverse, when the universe collapses into a fireball and then 'bounces' out in a rerun of the Big Bang, and does this repeatedly for ever. There are other strong theoretical reasons for believing that the universe will expand for ever, though ever more slowly. However, observations of very remote **supernovae** in other galaxies suggests that galaxies' speeds are not falling as fast as would be expected, and they may even be increasing because of a suspected repelling force between them that grows stronger with distance (see **cosmological constant**).

If the universe is to expand for ever, a dark, cold eternity lies ahead. The faintest, slowest-burning stars will fade out trillions of years from now. The dead stars will settle into supermassive **black holes** at the centres of the dead galaxies. After unimaginably vast periods, matter itself will decay. Even the **proton**, may decay after 10^{35} years, though this is not certain. The universe will become a dark sea of low-energy radiation, **electrons** and positrons. However, **parallel universes** undetectable by us may continue to be born and to evolve.

Future of the universe*

Years from now	Event
10^{10}	death of the Sun
10^{14}	death of the lowest-mass, slowest-burning stars
10^{18}	galaxies collapse into central black holes
10^{35}	protons decay (possibly)
10^{68}	solar-mass black holes evaporate
10^{100}	giant black holes evaporate

*This table assumes that the universe will expand for ever, rather than slowing down and then collapsing; and that the proton is not stable.

Uranus

The seventh **planet** from the Sun, the first new planet to be discovered (by William **Herschel** in 1781) after the invention of the **telescope**.

- Uranus is twice as far out as the sixth planet, **Saturn**.

- For half of its 'year', one pole of Uranus points towards the Sun; for the other half, it's the other pole's turn. This results in extreme seasons.

- The deep atmosphere is composed mainly of **hydrogen** and **helium**.

- The rotation rate of the **atmosphere** varies with **latitude**, from about 16 hours in mid-southern latitudes to longer than 17 hours at the equator.

- The space probe *Voyager 2* found 10 small moons in addition to the 5 visible from Earth.

SPIN DOCTORING

If the north pole of Uranus is taken to be the one that rotates counterclockwise when viewed from above (like the Earth's), then it lies *below* the plane of the planet's **orbit** – that is, the spin axis of Uranus is tilted at 98°. However, some astronomers prefer to call the other pole the north pole – which technically means that Uranus spins from east to west. This is the opposite of the other planets, with the exception of **Venus** and **Pluto**.

The ring system

Eleven thin rings around the planet's equator were discovered in 1977. The ring farthest from the planet centre (51,000 km/31,800 mi), Epsilon, is 100 km/62 mi at its widest point. In 1995, US astronomers determined the ring particles contained long-chain hydrocarbons. Looking at the brightest region of Epsilon, they were also able to calculate the **precession** of Uranus as 264 days, the fastest known precession in the **Solar System**. The rings are charcoal-black, and may be debris of former 'moonlets' that have broken up.

URANUS: STATISTICS

Average distance from Sun	Average distance from Sun (Earth = 1)	Orbital period (years)	Equatorial diameter	Mass (Earth = 1)	Tilt of equator
2.9 billion km/ 1.8 billion mi	19..2	84	50,800 km/ 31,600 mi	14.5	98°

Density (water = 1)	Escape velocity	Rotation period	Known satellites
1.27	22 kps/14 mps	17.2 hr	17

Ursa Major

The third-largest **constellation** in the sky, in the north polar region. Its seven brightest stars make up the familiar shape of the Plough (called the Big Dipper in the USA).

- The second star of the 'handle', called Mizar, has a companion star, Alcor.
- Two stars forming the far side of the 'bowl' act as pointers to the north pole star, **Polaris**.
- Many of the stars in Ursa Major are moving through space at the same speed and in the same direction. In addition, **Sirius** and several other stars in different parts of the sky share this common motion and so belong to the 'Ursa Major moving group'.

Ursa Minor

A small **constellation** of the northern hemisphere.

- It is shaped a little like the nearby Plough, which is part of **Ursa Major**.
- The bright north polar star **Polaris** lies at one end of the constellation.
- Two other bright stars in this group, Beta and Gamma Ursae Minoris, are called 'the guards' or 'the guardians of the pole'.
- The constellation also contains the orange subgiant Kochab, about 95 **light years** from Earth.

variable star

A star whose brightness changes, either regularly or irregularly, over a period ranging from a few hours to months or years. The different types of variability are closely related to different stages of stellar evolution.

- *Cepheid variables* regularly expand and contract in size every few days or weeks.

- Stars that change in size and brightness at less precise intervals include *long-period variables*, such as the red giant **Mira** in the **constellation Cetus** (**period** about 330 days), and *irregular variables*, such as some red **supergiants**.

- *Eruptive variables* emit sudden outbursts of light. Some suffer **flares** on their surfaces, while others, such as **novae** and Type I **supernovae**, result from transfer of gas between a close pair of stars.

- A *Type II supernova* is the explosive death of a star.

- In an *eclipsing binary*, the variation is due not to any change in the star itself, but to the periodic eclipse of a star by a close companion.

CHANGES AMONG THE STARS

Apart from a few novae, no star was observed to be variable until the end of the 16th century. Then a Dutch clergyman, David Fabricius (1564–1617), noticed the variability of **Mira** in 1596, and an English astronomer, John Goodricke (1764–1786) discovered the smaller but very regular variations of Delta Cephei in 1784. Fewer than 20 variable stars were known until the beginnings of **astrophotography** in the 1840s. More recently, electronic photometry has enabled the light curves of variables to be plotted very accurately.

Eclipsing binaries
Of 19,000 known variable stars, 80% are intrinsic variables and 20% are **eclipsing binaries** – pairs of stars in **orbit** around one another, whose combined light drops when one star eclipses the other. Much can be learnt of

the structure of the eclipsing system from detailed measurements of the light curve. In **spectroscopic binaries**, in which the velocities of both components can also be observed, it is possible to deduce the mass, diameter, and temperature of each star. The best known eclipsing binary is **Algol**. Some eclipsing binaries consist of pairs of stars orbiting very close to one another or almost in contact, and such stars show severe deformation due to their mutual gravitational attraction. Orbital periods of eclipsing binaries range from about 4 hours to 27 years.

Cepheid variables
A numerous class of variable stars of short to moderate period consists of **Cepheid variables**, named after the star Delta Cephei. This star pulsates in a very regular period of 5.37 days, with a light variation of 0.8 **magnitude**, making it just over twice as bright at maximum as at minimum. The periods of Cepheids range from 3 days to 50 days, and Henrietta **Leavitt** showed in 1912 that there is a relationship between their period and their brightness: the longer the period, the brighter the Cepheid. Once the relationship had been established, the observed period of a very distant Cepheid could be used to determine its intrinsic **luminosity**, and hence, taking into account its apparent luminosity, to find its distance. Thus Cepheids became important in calculating the distance of remote **galaxies**.

This period–luminosity relationship had to be revised in 1952, which doubled the accepted distances of the galaxies. This was the direct result of the recognition that W Virginis and RR Lyrae variables did not belong to the Cepheids, as had been assumed. W Virginis stars are sometimes called Type II Cepheids, while RR Lyrae stars are Cepheid-like variables with periods that fall in the range 0.3 to 0.7 days. They are sometimes described as cluster-type variables, as many of them are found in **globular clusters**.

Vega or Alpha Lyrae
The brightest star in the **constellation Lyra** and the fifth-brightest star in the night sky. It is blue-white, with a true **luminosity** 50 times that of the Sun.

- Vega is 25 **light years** from Earth.
- In 1983 **IRAS,** the Infrared Astronomy Satellite, discovered a ring of **dust** around Vega, possibly a disc from which a **planetary system** is forming.
- As a result of **precession**, Vega will become the north polar star about AD 14,000.

See also: *Polaris.*

Vega mission

A pair of Soviet space probes launched towards **Venus** in December 1984, which later visited Halley's Comet. (The name '*Vega*' comes from two Russian words meaning 'Venus' and 'Halley'.)

- On passing Venus in June 1985, both spacecraft released landing modules. During the descent, each lander released a 3.4-m/11-ft sounding balloon that floated at a height of 54 km/34 mi in the **atmosphere** of Venus, travelling 11,000 km/6,800 mi in two days.

- Precise tracking of the two balloons by **radio telescopes** on Earth, using VLBI (very long-baseline **interferometry**) to obtain high-resolution images, allowed astronomers to study the atmospheric circulation of Venus.

Venera

A series of 16 Soviet space probes launched towards **Venus** in the period 1961–83.

Venera missions

Date	Mission no.	Notes
October 1967	4	Entered the atmosphere of the planet but was destroyed by the intense heat and pressure before it reached the surface.
May 1969	5, 6	Entered the atmosphere; but were destroyed.
December 1970	7	First spacecraft to land on Venus.
July 1972	8	Landed on Venus.
October 1975	9, 10	First pictures from the surface.
December 1978	11, 12	Atmospheric probes sampled atmosphere and recorded lightning flashes.
March 1982	13, 14	First soil analysis.
October 1983	15, 16	Arrived in orbit, made radar maps of the cloud-covered planet.

Venus

The second **planet** from the Sun, and the planet most similar to the Earth in size and mass. Venus rotates on its axis more slowly than any other planet.

- Venus can approach Earth to within 38 million km/24 million mi, closer than any other planet.

- The planet rotates from east to west, the opposite direction to the other planets except *Uranus* (technically) and **Pluto**.

- Venus has an ion-packed tail 45 million km/28 million mi in length that stretches away from the Sun and is caused by the bombardment of the ions in Venus's upper **atmosphere** by the **solar wind**. It was first discovered in the late 1970s but it was not until 1997 that the Solar Heliospheric Observatory (**SOHO**) revealed its immense length.

THE ATMOSPHERE OF VENUS

Venus is shrouded by clouds of sulphuric acid droplets that sweep across the planet from east to west every four days. The atmosphere is almost entirely carbon dioxide, which traps the Sun's heat by the **greenhouse effect** and raises the planet's surface temperature to 480°C/900°F, with an atmospheric pressure of 90 times that at the surface of the Earth.

The surface of Venus
The surface of Venus consists mainly of silicate rock and may have an interior structure similar to that of Earth: an iron–nickel core, a mantle composed of more mafic rocks (rocks made of one or more ferromagnesian, dark-coloured minerals), and a thin siliceous outer crust. The surface is dotted with deep impact **craters**. Some of Venus's **volcanoes** may still be active. The largest highland area is Aphrodite Terra near the equator, half the size of Africa. The highest mountains are on the northern highland region of Ishtar Terra, where the massif of Maxwell Montes rises to 10,600–m/35,000 ft above the average surface level. The highland areas on Venus were formed by volcanoes.

The exploration of Venus
The first artificial object to hit another planet was the Soviet probe *Venera 3*, which crashed on Venus on 1 March 1966. Later *Venera* probes parachuted down through the **atmosphere** and landed successfully on its

---

VENUS: STATISTICS

Average distance from Sun	Average distance from Sun (Earth = 1)	Orbital period	Equatorial diameter	Mass (Earth = 1)	Tilt of equator
108.2 million km/ 67.2 million mi	0.72	225 days	12,100 km/ 7,500 mi	0.82	177.3°

Density (water = 1)	Escape velocity	Rotation period	Known satellites
5.25	10.3 kps / 6.4 mps	243 days	0

surface, analysing surface material and sending back information and pictures. In December 1978 a US *Pioneer* Venus probe went into orbit around the planet and mapped most of its surface by radar, which penetrates clouds. In 1992 the US space probe *Magellan* mapped 99% of the planet's surface to a resolution of 100 m/330 ft.

Very Large Telescope (VLT)

Major **telescope** under construction by the **European Southern Observatory** at the Paranal Observatory, Atacama, Chile. It comprises four 8.2-m/323-in reflecting Unit Telescopes, completed in 2000. There are also several moving 1.8-m/71-in Auxiliary Telescopes, the light beams of which will be combined in the VLT Interferometer (VLTI). These features will make it the world's largest and most advanced optical **telescope**. The use of **interferometry** will give the VLT the ability to produce extremely sharp images and to record **light** from the faintest and most remote objects in the universe.

Vesta

The third-largest **asteroid** in the **Solar System**, discovered in 1807 by the German astronomer Heinrich Olbers (1758–1840).

- Vesta is 500 km/310 mi in diameter.

- It **orbits** the Sun at a distance of 353 million km/219 million mi, with a **period** (the time it takes to circle the Sun) of 3.63 years.

- It is the only asteroid that ever becomes bright enough to be seen without a **telescope**.

- In July 1999 the NASA space probe *Deep Space 1* photographed the small asteroid Braille. The **spectrum** of light reflected from Braille indicates that Braille has the same composition as Vesta and so was probably chipped off during a past impact.

- Vesta has a shallow **crater** at its south pole, indicating that a large chunk of rock has been dislodged at some point.

Viking probes

Two US uncrewed space missions to **Mars**, each one consisting of an orbiter and a lander. They were launched on 20 August and 9 September 1975.

- *Viking 1* carried life-detection labs and landed in the Chryse lowland area on 20 July 1976 for detailed research and photos. It transmitted colour pictures and analysed the soil.

- *Viking 2* was similar in set-up to *Viking 1*; it landed in Utopia on 3 September 1976.

- The life-detection experiments produced a flurry of excitement when soil samples treated with nutrients emitted a burst of oxygen – the kind of reaction that simple plants could produce. After long analysis, mission scientists concluded that an inorganic reaction was involved.

- Designed to work for 90 days, *Viking 1* operated for six and a half years, going silent in November 1982.

- *Viking 2* functioned for three and a half years.

    See also: *life beyond the Earth.*

Virgo

A **zodiacal constellation** of the northern hemisphere, the second-largest in the sky.

- Virgo is represented as a maiden holding an ear of wheat, marked by first-**magnitude Spica**, the constellation's brightest star.

- Virgo contains the nearest large **galaxy cluster**, 50 million **light years** away. It consists of about 3,000 galaxies centred on the giant **elliptical galaxy** M87.

- Also in Virgo is the nearest **quasar**, 3C 273, an estimated 1.5 billion light years away.

- The Sun passes through Virgo from late September to the end of October.

- In astrology, the dates for Virgo are between about 23 August and 22 September (see **precession**).

VLA (Very Large Array)

The largest and most complex single-site **radio telescope** in the world. It is located on the Plains of San Augustine, 80 km/50 mi west of Socorro, New Mexico.

- It consists of 27 dish antennae, each 25 m/82 ft in diameter, arranged along three equally spaced arms forming a Y-shaped array.

- Two of the arms are 21 km/13 mi long, and the third, to the north, is 19 km/11.8 mi long.

- The dishes are mounted on railway tracks enabling the configuration and size of the array to be altered as required.

- Pairs of dishes can also be used as separate interferometers (see **interferometry**), each dish having its own individual receivers that are remotely controlled, enabling many different frequencies to be studied.

    See also: *radio astronomy.*

Volcano

An opening in a **planet's** crust through which hot fluids well up. On Earth, a volcanic mountain, usually cone-shaped with a **crater** on top, is formed around the opening, or vent, by the build-up of solidified lava and ashes (rock fragments). Most volcanoes arise on plate margins, where the movements of plates generate magma (molten rock) or allow it to rise from the mantle beneath. However, a number are found far from plate-margin activity, on `hot spots' where the Earth's crust is thin. There are more than 1,300 potentially active volcanoes on Earth.

Vulcanism has helped shape other members of the **Solar System**. In the case of the larger worlds, such as the Moon, **Mars**, and **Venus**, the driving force has been internal heat left over from the body's formation. In the case of **Jupiter's** moon **Io**, the heat originates from stretching and squeezing by the gravitational filed of the huge parent planet.

Von Neumann probes

Devices proposed speculatively to overcome the vast times required for any conceivable form of **interstellar exploration**. The probes are named after the Hungarian-born US mathematician John Von Neumann (1903–1957), who studied the mathematics of self-replicating machines. A Von Neumann probe would be a robot craft that, after arriving in a **planetary system** and surveying it thoroughly, would build a replica of itself, using raw materials mined from a suitable **planet** or **asteroid**. The two probes would then travel to two different further systems, taking decades to get there. After completing their work there, there would be four probes to continue exploration. The probes would proliferate like viruses, and the volume of space explored would expand through the **Galaxy** at roughly the speed of a single probe. If the probes could average one-tenth the speed of **light** (that is, travel at 30,000 kps/18,600 mps – unimaginably high by today's standards), it would take only a million years to explore the whole Galaxy.

Vostok
The first Soviet crewed spacecraft, in operation from 1961 to 1963. The name is Russian for 'east'. *Vostok* was a metal sphere 2.3 m/7.5 ft in diameter, capable of carrying one cosmonaut. It made flights lasting up to five days. *Vostok* 1 carried the first person to travel in space, Yuri **Gagarin**.

Voyager probes
Two US space probes that flew past the outer **planets** and have now left the **Solar System**.

- *Voyager 1*, launched on 5 September 1977, passed **Jupiter** in March 1979, and reached **Saturn** in November 1980.

- *Voyager 2*, though launched earlier, on 20 August 1977, was on a slower trajectory that took it past Jupiter in July 1979, Saturn in August 1981, **Uranus** in January 1986, and **Neptune** in August 1989.

- Like the *Pioneer* probes, the *Voyagers* have left the Solar System. Their tasks now include helping scientists to locate the position of the heliopause, the boundary at which the **solar wind** gives way to **interstellar matter**.

- Both *Voyagers* carry specially coded long-playing records called *Sounds of Earth* for the benefit of any other civilizations that might find them.

Vulcan
A hypothetical **planet**, suggested in the 19th century in order to explain problems that astronomers were having with the **orbit** of **Mercury**, the planet closest to the Sun. Mercury's movements could not be completely accounted for by **Newton's** laws of gravity. A French astronomer, Urbain Leverrier (1811–1877), proposed in 1845 that the discrepancy could be explained by the gravitational attraction of an undiscovered planet, which he called Vulcan, orbiting within the orbit of Mercury, only 30 million km/18.6 million mi from the Sun. All attempts to find Vulcan failed, and Leverrier turned his attention to **Uranus**, which was also deviating from its predicted path. Again, he attributed the perturbation to another planet, this time beyond the orbit of Uranus, and predicted where in the sky it would be found. This prediction was crowned with the discovery of **Neptune**, but Vulcan remained undiscovered. The Mercury problem remained unsolved until 1915, when Albert **Einstein** showed that the discrepancies in the orbit were a consequence of **relativity** and no undiscovered planet was needed.

white dwarf

A small, hot star, the last stage in the life of a star such as the Sun. They are the shrunken remains of stars that have exhausted their internal energy supplies.

- White dwarfs make up 10% of the stars in the **Galaxy**; most have a mass 60% of that of the Sun, but only 1% of the Sun's diameter, similar in size to the Earth.

- Most have surface temperatures of 8,000°C/14,400°F or more, hotter than the Sun. Yet, being so small, a white dwarf's **luminosity** may be less than 1% of that of the Sun.

- White dwarfs consist of degenerate matter in which gravity has packed the **protons** and **electrons** together as tightly as is physically possible, so that a spoonful of it has a mass of several tonnes.

- White dwarfs shine by giving out stored heat, as they slowly cool and fade over billions of years.

See also: *black dwarf.*

wormholes

Hypothetical 'tunnels' through **space-time** that could link the interiors of **black holes**. Anything travelling along a wormhole would seem to the outside observer to disappear at one place, and reappear at another, possibly at some time in the past or future. Whether they exist is uncertain, and whether anything could survive the stresses that would exist inside one is even more so.

See also: *time travel.*

X-ray

Electromagnetic radiation occupying the wavelength range 0.01 nanometre to 1 nanometre (1 nanometre, is one-billionth of a metre). X-rays lie between **gamma rays** and **ultraviolet radiation**. Very high-energy processes are required to produce them. **X-ray astronomy** is a fast-developing field.

X-ray astronomy

The study of **X-rays** from the universe, originating in intensely hot gas in interstellar and intergalactic space. Such X-rays are prevented from reaching the Earth's surface by the **atmosphere**, so detectors must be placed in **rockets** and **satellites**. The first celestial X-ray source, Scorpius X-1, was discovered by a rocket flight in 1962. Since 1970, special **satellites** have been put into orbit to **study** X-rays from the Sun, stars, and **galaxies**. Many X-ray sources are believed to be gas falling on to **neutron** stars or **black holes**.

X-ray telescope

An instrument designed to detect electromagnetic waves in the **X-ray** part of the **spectrum** and form images from them. X-rays cannot be focused by lenses or **mirrors** in the same way as visible **light**, and a variety of alternative techniques is used to form images. Because X-rays cannot penetrate the Earth's **atmosphere**, X-ray telescopes are mounted on **satellites**, **rockets**, or high-flying balloons.

year

Any of several different units of **time measurement** based on the movement of the Earth around the Sun.

- The *tropical* year is the period from one spring **equinox** to the next, and lasts 365.24219 days. It governs the occurrence of the seasons, and is the period on which the calendar year is based.

- The *sidereal year* is the time taken for the Earth to complete one **orbit** relative to the stars, and lasts 365.25636 days (about 20 minutes longer than a tropical year). The difference is the effect of **precession**, which slowly moves the position of the equinoxes.

- The anomalistic year is the time taken by any **planet** in making one complete revolution from **perihelion** to perihelion; for the Earth this period is about five minutes longer than the sidereal year, because the perihelion shifts, owing to the gravitational pull of the other planets.

- The *calendar year* consists of 365 days, with an extra day added at the end of February each leap year.

- Leap years occur in every year that is divisible by 4, except that a century year is not a leap year unless it is divisible by 400. The rule makes 2000 a leap year, but 2100 will not be.
  See also: *day.*

zenith
The point on the **celestial sphere** directly above the observer; the **nadir** is the point below, diametrically opposite.

zodiac
Band around the heavens containing the **ecliptic** (the yearly path of the Sun) and the paths of the Moon and **planets**. When it was devised by the ancient Greeks, only five planets were known, making the zodiac about 16° wide. In astrology, the zodiac is divided into 12 signs, each 30° in extent. These do not cover the same areas of sky as the astronomical **constellations**. The 12 astronomical constellations are uneven in size and do not between them cover the whole zodiac, or even the line of the ecliptic, much of which lies in the constellation of **Ophiuchus**.

The workings of the zodiac
The word 'zodiac' is of Greek origin, but the idea of such a zone is more ancient. The sequence of the signs is eastwards, following the motions of the Sun and Moon through the constellations, and is regarded as beginning at the point that marks the position of the Sun at the time of the March **equinox**. This point is sometimes called the **First Point of Aries**. The equinoxes occur as the Sun enters the signs of **Aries** and **Libra**, at which times the Sun passes directly above the **equator**. The **solstices** occur as the Sun enters the signs of **Cancer** and **Capricorn** and is passing directly over the tropic of Cancer and tropic of Capricorn, which are circles of **latitude** 23° 27' north and south of the equator.

Because of **precession** the equinoctial point, and with it the zodiacal signs, moves westwards through the constellations at a rate of one sign in 2,150 years. The spring equinox, which is now in the constellation **Pisces**, would at the time of **Hipparchus** have been in the constellation **Aries**. At the time the present constellation groupings were first made, 2,000 years earlier, the spring equinox would have been in **Taurus** and the beginning of the zodiac would have been marked by the Pleiades star cluster.

Appendix

Astronomy: chronology

2300 BC	Chinese astronomers make their earliest observations.
2000 BC	Babylonian priests make their first observational records.
1900 BC	Stonehenge is constructed: first phase.
434 BC	Anaxagoras claims the Sun is made up of hot rock.
365 BC	The Chinese observe the satellites of Jupiter with the naked eye.
3rd century BC	Aristarchus argues that the Sun is the centre of the Solar System. Eratosthenes accurately measures the size of the Earth.
2nd century BC	Hipparchus draws up accurate catalogue of about 1,000 stars, and discovers precession of equinoxes.
2nd century AD	Ptolemy proposes his Earth-centred system, which dominates the astronomy of the Middle Ages.
4th–8th century AD	In pre-Columbian America, the Maya develop an intricate and accurate calendar.
1543	Copernicus publishes his Sun-centred system of the universe.
1577	Tycho Brahe, greatest of pre-telescopic observers, observes a comet and concludes it wanders among the planets.
1608	Hans Lippershey invents the telescope, which is used independently by Galileo Galilei and by Simon Marius to observe the sky.
1609	Johannes Kepler's first two laws of planetary motion are published (the third appears in 1619).
1632	Europe's first official observatory is established in Leiden in the Netherlands.
1633	Galileo's theories are condemned by the Inquisition.
1675	The Royal Greenwich Observatory is founded in London, UK.
1687	Isaac Newton's *Principia* is published, which includes his 'law of universal gravitation'.
1705	Edmond Halley correctly predicts that the comet that had passed the Earth in 1682 will return in 1758; the comet is later known by his name.
1781	William Herschel discovers Uranus and recognizes stellar systems beyond our Galaxy.
1801	Giuseppe Piazzi discovers the first asteroid, Ceres.
1814	Joseph von Fraunhofer first studies absorption lines in the solar spectrum.
1846	Neptune is identified by Johann Galle, following predictions by John Adams and Urbain Leverrier.
1859	Gustav Kirchhoff explains dark lines in the Sun's spectrum.

1887	The earliest photographic star charts are produced.
1889	Edward Barnard takes the first photographs of the Milky Way.
1923	Edwin Hubble proves that the galaxies are systems independent of the Milky Way, and by 1930 has confirmed the concept of an expanding universe.
1930	The planet Pluto is discovered by Clyde Tombaugh at the Lowell Observatory, Arizona.
1931	Karl Jansky founds radio astronomy.
1945	A radar pulse is reflected from the Moon by Z Bay of Hungary and the US Army Signal Corps Laboratory.
1948	The 5-m/200-in Hale reflector telescope is installed at Mount Palomar, California.
1957	The Jodrell Bank telescope dish in the UK is completed. The first Sputnik satellite (USSR) opens the age of space observation.
1962	The first X-ray source is discovered in Scorpius.
1963	The first quasar is discovered.
1967	The first pulsar is discovered by Jocelyn Bell and Antony Hewish.
1969	The first crewed Moon landing is made by US astronauts.
1976	A 6-m/240-in reflector telescope is installed at Mount Semirodniki, USSR.
1977	Uranus is discovered to have rings. The spacecraft *Voyager 1* and *2* are launched, passing Jupiter and Saturn 1979–81.
1978	The spacecraft Pioneer *Venus 1* and *2* reach Venus. A satellite of Pluto, Charon, is discovered by James Christy of the US Naval Observatory.
1986	Halley's comet returns and is studied by a fleet of uncrewed space probes. *Voyager 2* flies past Uranus and discovers six new moons.
1987	Supernova SN1987A flares up, becoming the first supernova to be visible to the naked eye since 1604.
1989	*Voyager 2* flies by Neptune and discovers eight moons and three rings.
1990	The Hubble Space Telescope is launched into orbit by the US space shuttle.
1991	The space probe *Galileo* flies past the asteroid Gaspra, approaching it to within 26,000 km/16,200 mi.
1992	The COBE satellite detects ripples from the Big Bang that mark the first stage in the formation of galaxies.
1994	Fragments of comet Shoemaker-Levy 9 strike Jupiter.
1996	US astronomers discover the most distant galaxy so far detected. It is in the constellation Virgo and is 14 billion light years from Earth.
1997	The satellite Hipparcos measures the distances to many nearby stars with unprecedented accuracy, helping to improve estimates of the age of the universe. Two new satellites are discovered circling Uranus, bringing the total number of its moons up to 17.
1998	NASA announces the discovery of up to 300 million tonnes of ice on the surface of the Moon. The ice exists as a thin layer of crystals inside some craters that are permanently in shadow.
1999	Observations of remote supernovae suggest that the expansion of the universe may be accelerating.

The Brightest Stars

Scientific name	Common name	Distance from Sun (light years)	Apparent magnitude*	Absolute magnitude*
Alpha Canis Majoris	Sirius	9	−1.46	1.4
Alpha Carinae	Canopus	1,170	−0.72	−2.5
Alpha Centauri	Rigil Kent	4	−0.27	4.4
Alpha Boötis	Arcturus	36	−0.04	0.2
Alpha Lyrae	Vega	26	0.03	0.6
Alpha Aurigae	Capella	42	0.08	0.4
Beta Orionis	Rigel	910	0.12	−8.1
Alpha Canis Minoris	Procyon	11	0.38	2.6
Alpha Eridani	Achernar	85	0.46	−1.3
Alpha Orionis	Betelgeuse	310	0.50**	−7.2
Beta Centauri	Hadar	460	0.61**	−4.4
Alpha Crucis	Acrux	360	0.76	−4.6
Alpha Aquilae	Altair	17	0.77	2.3
Alpha Tauri	Aldebaran	25	0.85**	−0.3
Alpha Scorpii	Antares	330	0.96**	−5.2
Alpha Virginis	Spica	260	0.98**	−3.2
Beta Geminorum	Pollux	36	1.14	0.7
Alpha Piscis Austrini	Fomalhaut	22	1.16	2.0
Beta Crucis	Mimosa	420	1.25**	−4.7
Alpha Cygni	Deneb	1,830	1.25	−7.2

* A star's brightness is referred to as its 'magnitude'. 'Apparent magnitude' is brightness as seen from Earth. 'Absolute magnitude' is measured at a standard distance of 32.6 light years or 10 parsecs from the star.
** Variable star.